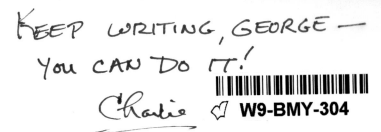

KEEP WRITING, GEORGE —
You CAN DO IT!
Charlie

Charles Hig
12/08

W ar is an ugly thing, but not the ugliest of things. The decayed and degraded state of moral and patriotic feeling which thinks that nothing is worth war is much worse. The person who has nothing for which he is willing to fight, nothing which is more important than his own personal safety, is a miserable creature and has no chance of being free unless made and kept so by the exertions of better men than himself. *- John Stuart Mill – English economist & philosopher (1806 – 1873)*

We Were
Crewdogs
IV

We Had
To Be Tough

**Edited by
Tommy Towery**

Should the decision be made to publish a future volume of stories such as this and you want to participate, please contact the editor to let your desire be known or visit our web site at:

www.wewerecrewdogs.com

Contact Info:
Tommy Towery
5709 Pecan Trace
Memphis, TN 38135
ttowery@memphis.edu

1000 Hour B-52 Pin
Available for purchase from www.wewerecrewdogs.com

Dedication

George Golding in the cockpit.

This volume is dedicated to Colonel George Golding, USAF (Ret.). George passed away on November 15, 2007 only days before he could see the efforts of his contribution to *"We Were Crewdogs III – Peace Was Our Profession"*. According to his wife, Sue, "He put up a valiant fight and never lost his wonderful sense of humor throughout the ordeal. He did pass peacefully and I am sure he is now in heaven telling many more stories with all of the warrior airmen that went before him."

George was a vital contributor and supporter to this series of books and we are happy that we were able to preserve some of his great B-52 stories for his friends and family. In addition to his stories, his contribution of many pictures from his collection of photos was invaluable in illustrating several other stories.

i

Table of Contents

Chapter 3 - Southeast Asia

Chapter 4 - Lest We Forget

Résumé

Tough [tuhf] - *adjective –* to be capable of great endurance; sturdy; hardy.

Foreword
E.G. "Buck" Shuler, Jr.
Lt. Gen., USAF, Retired

My immersion in the marvelous Crewdog series in Volumes I, II, and III, caused me to reflect on the significance and meaning of these wonderful stories, which continue to be related in this volume.

"We Were Crewdogs" brought back many significant memories of my 23 years of service in SAC, of similar experiences, of people and places I have known, and the Cold War we won. I thought about the millions of hours flown by bomber, tanker, and reconnaissance crews ranging over the entire planet and often under difficult conditions of weather and emergencies. I thought about the nearly 2,500 Crewdogs who lost their lives flying training and operational combat missions, and most importantly their surviving family members. I thought about our sturdy aircraft, the B-50, B-47, B-52, FB-111, KC-97, KC-135, KC-10, U-2, SR-71, RC-135, EC-135, and E-4B as well as all of the support aircraft assigned to SAC. I thought about the thousands upon thousands of both ground and airborne alert duty days, when the norm was two seven-day ground alert tours per month in addition to flying three to four 10-12 hour training missions as well as 24-25 hour airborne alert missions.

I thought about the continuing stream of unit evaluations, ORIs, Bar and Buy Nones, CEVGs, standboard rides, instrument checks, command control testing, SIOP target study and crew certifications, and on and on and on! I thought about the various world events and crises Crewdogs were involved in, the first and second Berlin Crises, the Cuban Missile Crisis, operations Bullet Shot, Arc Light, Young Tiger, Line Backer I and II, and many, many more.

I thought about our great leaders, Generals George C. Kenney, Curtis E. LeMay, Thomas S. Powers, John D. Ryan, Joseph J. Nazzaro, Bruce K. Holloway, John C. Meyer, Russell E. Dougherty, Richard H. Ellis, Bennie L. Davis and others. I thought about SAC's humble beginnings at Andrews AFB, Maryland, evolving out of the WWII strategic bombing experience, to its peak strength of 248,000 men and women in 1962 with headquarters at Offutt AFB, Nebraska, to its sad deactivation on 1 June 1992, exactly one year after I retired. Peace Was Our Profession and SAC was the biggest and best war fighting command ever assembled, and the Crewdogs made it so! Deterrence worked! The Soviet Union and the Warsaw Pact nations were defeated and again the Crewdogs made it happen!

I am honored to have been invited to pen this foreword for I will always be a Crewdog at heart. Tommy Towery, in particular, and his colleagues are to be commended for the compilation of the Crewdog volumes, which serve to preserve the history and heritage of SAC from March of 1946 to June of 1992. It is my sincere hope that all readers will read these inspirational stories and come away with a deep seated appreciation for what the SAC Crewdogs accomplished and sacrificed for the long term safety and well being of our nation. Americans should be eternally grateful!

Chapter One

Military Career [mil-i-ter-ee] [kuh-reer] – *noun* - An occupation or profession, especially one requiring special training, for, or pertaining to war - followed as one's lifework.

Edward O. Martin

Peace is Our Profession – But Why?
Bill Fritz

"Peace is our Profession" was the proud motto of the Strategic Air Command. Did you ever wonder how, when, where, and who originated this motto under which so many of us proudly served? I was, and after a visit to capture some other BUFF data, asked the assistance of the Air Force Historical Research Agency that is located at the Air Force Research Library at Maxwell Air Force Base. Somewhat to my surprise, this is what I discovered.

As the original members of the Strategic Air Command will know, and those of us who joined the Command later may not, the original motto of SAC was "War is our profession – Peace is our product." That was established at Offutt Air Force Base in March of 1946. It was reported in an article for the B-52 Stratofortress Association news letter in the November 2005 issue by Major General Chris Adams (USAF ret.), that the name was changed during the winter of 1957 during a personnel initiative to improve enlisted personnel retention.

There was to be a large Christmas tree in front of the Headquarters building fondly known as "Building 500" or "Ground Zero" by many. The tree was the 'Tree of Peace," and for each reenlistment a light on the tree was to be illuminated and as unit reenlistment goals were met, the names of the commanders were to be added to the status board adjacent to the tree. The problem was that the intended motto, "Maintaining Peace is our Profession" did not fit. Both Maj. Gen. Adams and Alwyn T. Lloyd, in his book *A Cold War Legacy: A tribute to Strategic Air Command*, are in agreement that the Project Officers for the tree were responsible for the change to the motto that most of us recognize. The intended motto was shortened to "Peace is our Profession." Lloyd names Lt. Col. Edward O. Martin, the Director of Personnel at HQSAC, and CWO Ben Kohot as the Project Officers responsible for this change.

Both Maj. Gen. Adams and Lloyd are also in agreement that Col. Van Vliet, Eighth Air Force Director of Public Information, took the idea back to Westover AFB and placed the motto on a sign at the main gate. Other 8th AF SAC units followed suit. The popularity was so positive that SAC achieved a reenlistment record that year, and the official change in the SAC motto was made in 1958. Not so widely know was that Lt. Col. Martin was already using this motto to gain support for the Headquarters in briefings to Omaha civic organizations that made contributions to fund some of the recreational facilities that Offutt needed. This information was obtained in personal conversations with Brig. Gen. Martin who was later the Wing Commander of the 424th BMW at Columbus AFB, and the 2nd BMW at Barksdale AFB. He was later assigned to the Joint Chiefs of Staff, and retired as Commander 42nd AD at McCoy AFB in 1974

There is more. In paperwork provided by the AF Historical Research Agency, on 11 Jan 1990, in correspondence from Maj. Gen. Donald L. Marks, Chief of Staff HQ/SAC, to HQ USAFHRC/RS, the motto was actually changed BACK to "War is our Profession – Peace is our Product" The rational contained in that correspondence is stated as self-evident: "Readiness for war and the willingness to engage in it when called upon is the profession of the warrior, a profession which deters aggression and preserves peace."

That is not the end. In a letter on 13 August 1991, Arlen D. Jameson, Chief of Staff, HQ SAC, wrote to HQ USAFHRC/CC stating that "CINCSAC has decided to restore the time-honored motto which served Strategic Air Command so well for almost four decades" and elaborated that the then current motto "…was not well received" and that "…the time honored and traditional Peace ….is our Profession is the overwhelming preference of SAC people." General McPeak was advised of this final change on 6 September, 1991 and the AF Historical Agency officially recognized the final change on 12 September 1991. The motto "Peace … is our profession" remained the motto until the disestablishment of The Strategic Air Command on 1 June 1992.

Peace through the Generations
Karl Nedela

Twenty years of my Air Force career was dedicated to living up to the motto of "Peace is Our Profession". I found that when in a position to be shot at or shot down, "Peace" is still the number one item on the agenda - but sometimes you had to fight for it.

What did it mean to me to be a Crewdog trying to keep the peace? When I start to put words on paper, the "little gray cells" activate themselves and wonderful events and interesting people come to mind. While I had many things that I can recall about my B-52 days, I will never forget one of the paths that led me up that road.

In the summer of 1953, before my B-52 days, I flew my first mission as a gunner in a B-36 at Carswell AFB, Texas. The crew with which I was training was headed to Limestone AFB, Maine. Many may be unfamiliar with the base by that name. A year later, in 1954, the name was changed to Loring AFB, in honor of Maj. Charles Joseph Loring, Jr. He was a native of Portland, Maine, and was awarded the Medal of Honor posthumously, after deliberately crashing his crippled F-80 fighter into a gun emplacement on Sniper Ridge during the Korean War. I remember marching in the parade for the dedication of the base.

Peace, like war, encompasses many generations. If you make a career in the military, odds are that you will be involved in more than one conflict. Just like humans, aircraft also bridge one conflict to another. Maj. Loring had been a fighter pilot in World War II and was a POW in that war, before being killed in Korea in 1952. In 1953 when I made that first flight, World War II had been over eight years (not a long time in the history of things) and the Korean Conflict was drawing to an end. We would later find that World War II aircraft and personnel would both still be around even in the Vietnam conflict.

I had a personal encounter with one such memorable World War II person. While severing on my B-36 crew I had the honor to fly with our 2nd Engineer, MSgt Theodore "Jake" Laban. Just as I transitioned from one aircraft to another in my career, so had he. Jake had been a flight engineer on a B-25 that participated in the attack against Japan in April of 1942. That was the famous "Doolittle Raid" that had originated from the deck of a Navy carrier off the coast of Japan. Jake's plane was blown far off course after the bombing of Tokyo and he and his crew ended up in Siberia, USSR. There they were interned for an extended time and later escaped. The experiences of this crew would make a first class movie. Today, very few people have even heard of these events.

Whenever we asked Jake about his life there, his answer was always, "Read the book!" The Loring AFB library actually had a copy of the book of which he spoke and I did read it. It was titled *"Guests of the Kremlin"* and was written in 1949 by Lt. Col. Robert G. Emmens, the copilot on his crew.

I have several copies of the book now and over the years have shared them with my family. It is one of the most fascinating and interesting books I have ever read. The book tells of being transported thousands of miles in the USSR and the dismal living conditions of the people in that "slave nation". If anyone got too friendly with the crew, they disappeared - never to be seen again by those Americans.

The weather and living conditions they survived shows the wonderful endurance of the American Crewdog during World War II and the tradition that has endured up to the present.

Being a Crewdog for 20 years helped to enrich my life. Throughout my years I had the honor to serve with so many talented and dedicated people. Theodore "Jake" Laban was just one of those "Crewdogs".

After flying my first mission as a B-36 gunner at Carswell AFB, Texas, I think it was ironic that exactly 20 years later, I flew my last mission in a B-52 from the same base.

I eventually retired from the service to my country and my family and I headed to Killeen, Texas. I proudly served 23 years in the Air Force and then proudly served 23 years teaching and coaching in the Killeen Independent School District. Just as I remember Jake, perhaps someday other Crewdogs will still brag that they once flew on a B-52 crew with a Crewdog like me that had actually flown bombing missions in Vietnam. Life goes on.

Glasgow Crew E-09 (L to R) AC -Bob Stephenson, CP - Cliff Hamby, RN - Dave Lowell, Nav - Murray Newton, EW - Marv Howell, Guns - Ithamer Armfield.

The Real Veterans on the Crew Force
Marv Howell

When I arrived in the 322nd Bombardment Squadron in 1964, a quick look around showed that there was a wide range of ages and ranks in the B-52 crew force. I am sure it was not unusual, but up to that time most of my Air Force contacts had been with captains and lieutenants, so it was different to discover that Glasgow AFB, Montana, had a lot of majors and lieutenant colonels.

In 1964 the B-52 had been in service less than 10 years, so unlike how things would be in later years, many of the existing Crewdogs did not start and finish their careers in the mighty BUFF. There were several who had not done so, and many who had not even started their flying careers in the B-47, its predecessor.

One of those was the Squadron Commander at the time, Colonel Leroy Fassman. He didn't spend a lot of time with the crews but could usually be found in the Ops Office that was located inside the Alert Facility. He worked a lot with the senior Aircraft Commanders (A/Cs) but only from time-to-time would he associate with the lowly other crew members. After I had been in the squadron for a while, he came into the dining room for coffee and we had the occasion to talk about his WWII experiences. During our conversation I found out that he had flown the P-39 Aircobra in the Pacific Theatre. Since I was an avid model-builder at that time, I asked him if he had any pictures of his airplane. Sometime later he brought me a faded shot of a P-39 with a big Cobra painted on the side-door. Yes, the P-39 had doors! Fortunately there was a decal available that would serve the purpose on a model kit. It took a couple of weeks for me to crank out a suitable model using the Monogram 1/48th scale kit as the basis of the project. When I gave it to him, he took the time to share with me a few more stories from the island war in the 1940s. I changed my appreciation of him significantly after hearing those tales.

Lt. Colonel Wallace Yancey was the Squadron Ops Officer who was frequently a gadfly and well known for calling most of the Crewdogs "Ace." He was a regular fixture in the dining room and in the mission planning rooms. For some reason, for which I never knew, my crew was one of his favorites to spend time with. The conversation would usually begin with "Well Ace, what have you done for the good of the Air Force today?" That would be followed with whatever topic he wanted to discuss, and it seemed like he always found some item to discuss. One time his topic of choice was chewing me out for some error on a mission or planning or something he had heard about. I protested, insisting that he was riding me too much. His reply was "Well Ace...if the shoe fits!"

I replied to that statement with, "Colonel, it doesn't always fit; you just keep trying to cram it on my foot." He laughed at that comment but it didn't change anything in his future visits.

In spite of his demeanor, everyone liked him. Once in a lighter moment, he revealed he had flown as a crewmember on the B-45 Tornado. Someone else added the fact that he not only flew it, but in 1952, as a Captain, he won the MacKay Trophy for the first non-stop flight across the Pacific Ocean in the B-45. That trophy is awarded for the "most meritorious flight of the year" by an Air Force person, persons, or organization.

Another interesting person was, Lt Col Al Hodkinson - my first A/C. I never learned much about his prior Air Force experience but he was a great leader. He was a stickler for getting the job done right and served as a positive motivator for the whole crew. He would sit with each of us individually and discuss how we could improve our performance in our specialty and as officers. He gave us great respect, regardless of rank, and we reflected it back to him. He also served as a great buffer between our crew and the staff and did the same for the whole alert force. He was usually the Senior Aircraft Commander on alert and wouldn't hesitate to 'correct' any errant directive or restrictions imposed on the alert pad. I later would apply a lot of the principles of leadership I learned from Colonel Hodkinson - mostly related to how to look out after subordinates. Serving on his crew made my start as a Crewdog a very positive experience. I wasn't on his crew very long, but it left a positive impression on me.

Lt Col "CY" Cykoski was a contemporary of Al Hodkinson. In fact, they were separated by only a couple of days in their dates-of-rank. That issue frequently came up when the Senior A/C on alert was designated. Cy was somewhat less vocal than Al but very conscious of who was "in-charge" on alert. He was not one of the poker-playing, TV-watching group members but nonetheless, his presence on alert was felt by everyone.

Lt Colonel Harry Newsome was another one of the leaders in the squadron. Although I never served on his crew, I spent a lot of time talking with him. We later wound up together at Dyess AFB, Texas, when Glasgow closed. He became a 'wing-weenie' at Dyess so we didn't have as much contact there. During our first Arc Light tour his crew had the quarters next to ours on the third floor of the BOQ. He was building a large Model of a Chris-Craft cruiser from balsa wood and I shared an interest as a modeler. That led to many long conversations between flights. That was how I found out that he had flown P-38s in the Mediterranean during WWII. He had two confirmed kills and one probable kill while flying with the 1st Fighter Group. In a letter to me later he commented that ground kills were not counted at that time in the war and he had several. His photo appears twice in the book *"An Escort of P-38s - The 1st fighter Group in World War II"* by John D. Mullens, Phalanx, 1995.

At Glasgow I built a model for him of the P-38 he flew. It took some modifications to the Monogram P-38J kit to back date it to the same version he flew, but I did it. After he retired, he moved to Little Rock, Arkansas, and I sent him a copy of the book I had found. He was in the process of moving to Michigan but still took the time to write me a thank-you letter. I have since lost contact, but not the memory, of a true veteran of WWII.

Sergeant Hance Storus was also a type of SAC pioneer. He was a gunner at Glasgow but I discovered he had been one of the early enlisted Electronic Warfare Officers. He was a pretty quiet guy and didn't talk a lot about his service as an EW. Like others of the time, he lost his job when the AF decided that EWs should be officers and not enlisted personnel. I don't know much about his early service and we went separate ways when Glasgow closed.

So these are a few of the outstanding officers and true combat veterans I knew at Glasgow. I am sure there are more; I just didn't hear their stories. The standard line was that the "Ninety-worst" was one of the best wings in SAC. I didn't think so at the time, but looking back I would have to agree.

My later experiences didn't match up to that first assignment and the "real" veterans I had the privilege to serve with in the 91st Bombardment Wing and the 322nd Bombardment Squadron.

Wurtsmith Crew E-38, taken 1n 1968, just prior to rotating through Arc Light. (L-R) CP - Charlie Davis, EW - Ron "Moose" Artman, Sq Cmdr Lt Col Goyne, N - Art Thompson, AC - Jim Reiman, G -Unknown, RN - John Burris.

Draft Dodger to Crewdog
Art Thompson

This is my story of how I went from an army Draft Dodger to a BUFF Crewdog in three short years. No, I didn't run off to Canada, or demonstrate against the war or even burn my draft card. I simply could not see myself being drafted into the army, which is what would have happened had I done nothing.

The year was 1963. JFK was assassinated, the Beatles were about to launch the British invasion, and the Vietnam War was heating up. Having graduated from Purdue University and being single, my draft status suddenly changed from 4F to 1A, meaning I went from college deferral status to draft status. I knew my status would change as soon as I graduated, but I was in denial. I wasn't running from the military - I was running from the army. After weighing my options of Navy Supply Officer on board ship (OCS), and Air Force Nav School (OTS), I chose the Air Force. At that time the waiting list for Air Force pilot training was two years, and my draft board had no sense of humor. All the while this selection process was going on, I couldn't help thinking about stories of young kids dreaming of an Air Force flying career ever since they were five years of age, and here I was on the threshold of an Air Force flying career as a way of avoiding being drafted into the army. It was one of the best decisions I ever made.

So, it was three months (90 day wonder) of Officer Training School (OTS) at Lackland AFB, Texas, 49 weeks of Undergraduate Nav School at James Connally AFB in Waco, Texas, six months of Nav/ Bomb Training at Mather AFB, California, and three months of B-52 Combat Crew Training at Castle AFB, California. Notice, there is no mention of Survival School. More on that later. Guess I was one of those geeks at Nav School who enjoyed Celestial Grid Navigation, allowing me to graduate relatively high in my class. That, in turn, allowed me a good shot at getting my first pick of operational assignments which was Wurtsmith AFB, Michigan, home of the B-52H. With the Vietnam War looming on the horizon, the selection choices were limited to location. With the exception of a few KC-135 slots and a Nav School Instructor slot, it was a given that almost everyone in the class was going to Crewdog School.

The year was 1966. LBJ was president, Bob Dylan was singing protest songs, and the BUFFs had begun flying 12-hour "ball-busters" out of Guam. In the mean time, I reported to Wurtsmith, my first operational assignment, in mid April. I subsequently certified the EWO Mission June 1, and went on alert June 1. Do you get the impression they were glad to see me at Wurtsmith? I quickly realized I was another "warm body" in the SAC scheme of things. One interesting side bar to my experience was that I became EWO certified and on alert, and still had not attended Survival School. My records indicated that I had a "waiver" due to the need to send "high priority" personnel prior to rotating through Southeast Asia. More on that later.

It seemed like I was on alert more than not. The normal rotation was every third week, but there were several back-to-back tours that I remember. I was the Nav on crew E-38 for most of my three-year Wurtsmith tour. Running and playing pool kept me sane on alert. Some people read, some played cards, some golfed on the putting green. I played so much pool that I finally broke down and bought my own stick. Not that I'm a pool shark, but the sticks on alert were all bowed and warped out of shape. Running was also therapeutic for me. I often would run the .7 mile service road that circled the alert Christmas tree, usually doing two laps. My time at Wurtsmith coincided with a research project that Air Force Dr. Ken Cooper was conducting

17

regarding aerobic training and how it might replace the existing annual 5BX physical conditioning test. Wurtsmith happened to be one of the Air Force Bases where he compiled data for the 1.5 mile run, which eventually did replace the 5BX test. I'll never forget the one day everyone on my crew had to run the 1.5 miles on alert. My EW was a heavy Pall Mall smoker, and about half way through the run he stopped, sat down on the pavement, and lit one. "Smoke 'em if ya got 'em."

The year was 1968. LBJ was still president, Jane Fonda was in bed with the Viet Cong, and the BUFFs were still flying 12-hour "ball-busters" out of Guam, with an occasional eight-hour mission out of Kadena and four-hour mission out of U-Tapao. My crew, E-38, was notified in June that we would be the third Wurtsmith crew to go to Arc Light. (Jim Reiman-P, Charlie Davis-CP, John Burris-RN, Art Thompson-N, Ron "Moose" Artman-EW, Gunner-Unknown). I don't remember the gunner's name because, for some reason, my crew had about three or four gunners rotate through at various times. There must have been lots of turmoil in the gunner ranks at that time.

We first rotated through Castle for two weeks of "D-difference" school, since we were an "H" Model Crew and Arc Light had "D" models. Ever since being notified in June we would be going to Arc Light, we kept hearing rumors that all Arc Light tours were being extended to six months, even though our orders indicated three months. As my pilot often said, "Stand by for a ram." I'll never forget arriving on Guam at 0200, only to be greeted by a colonel saying, "Welcome to Guam…you've been extended to six months." So there I was, sitting on Guam, about to commence flying B-52 combat missions over Vietnam, and sitting on a "waiver" to Air Force Survival School due to the need to send "high priority" personnel rotating through Southeast Asia. Long story short…I never did go to Survival School at any time during my 28-years of military service. The Air Force works in strange ways.

I personally flew 50 combat missions on that Arc Light tour. My crew was sometimes split up for various reasons, so I'm not sure how many the other guys flew. Most of our missions were the 12-hour "ball-busters" out of Guam, with three weeks of flying out of U-T and three weeks out of Kadena. I was always amazed that mission planning would schedule another couple of "practice" bomb runs over Vietnam, only to add to the misery of a long mission. I was fortunate in seeing a USO show with Bob Hope at U-T over Christmas. I also heard the third

Super Bowl over armed forces radio at U-T when Joe Namath surprised the odds makers by winning the game for the N.Y. Jets. I was DNIF four days at U-T with the dreaded GI's, the only time I was DNIF during my 28-year flying career. Must have been the ice cubes in the drink. The flight surgeon gave me some "liquid plug", and that took care of it. We had four R&Rs to the Philippines, Hong Kong, Bangkok, and CCK, with a few side trips to Sadahip. With the social temptations that go along with R&R in that part of the world, I'm happy to report I did not come down with the dreaded "Philippine Fall-Apart" or the "Hong Kong Dong." Some high-ranking colonel in SAC thought it would be a neat idea to schedule an R&R in Vietnam so we could see first hand the destructive power of the mighty BUFF. We politely declined the invitation. We had very little personal time: it was eat, sleep, and fly. The name of the game was "turn the airplanes and aircrews in minimum time with minimum maintenance." We purchased a '50 Ford for $100 that we quickly named the "Guam Bomb" when we first arrived. We subsequently sold it to an incoming crew for the same price when we left. There were lots of "Guam Bombs" changing hands.

We did the usual shopping that goes with Southeast Asia TDYs: Noratake China, silk ties, cashmere coats, teak wood Buddha's and elephants, Seiko watches, cameras, Sony real-to-real tape decks. We tried to have fun and see the humor whenever we could. My crew had a slogan that often popped up -"It ain't much of a war, but it's the only one we got."

After returning to Wurtsmith, I submitted my paperwork to separate from the Air Force. I had no idea what I was going to do for a career, but it was time to move on and try something else. Ironically, I subsequently hooked up with the Wisconsin Air National Guard as a Nav in KC-97's for seven years, followed by 15 years in KC-135's. My 22 years in the Guard were, by far, the most fun years. And to think on active duty in SAC we would make fun of the Guard guys because their gig lines were off and they shined their shoes with a Hershey bar and polished them with a brick. I soon discovered the Guard guys had the last laugh.

Back row: (L to R) Captain Earl J. Farney - radar navigator, Captain Ellie G. "Buck" Shuler, Jr. - pilot, Captain Arthur Craig Mizner - aircraft commander. Front row: (L to R) First Lieutenant Barry L. Gomborov - Electronic Warfare Officer, T/Sgt. James T. Starr - aerial gunner, First Lieutenant Norman D. Elder – navigator.

Crew of the Quarter –
Dyess E-15
Arthur Craig Mizner

The headlines on the *"Dyess Peacemaker"*, Volume III, Number 43, Abilene, Texas, of Friday, October 23, 1964 newspaper front page states "Bomber Crew E-15 Tops during Third Quarter."

Crew E-15 has been chosen as the 337th Bombardment Squadron, 96th Strategic Aerospace Wing's Crew of the Quarter for July, August, and September 1964.

Crew members are: Captains Arthur Craig Mizner, aircraft commander: Ellie G. "Buck" Shuler, Jr., pilot; and Earl J. Farney, radar navigator; First Lieutenants Norman D. Elder, navigator and Barry L. Gomborov, electronics warfare officer; and TSgt. James T. Starr, aerial gunner.

The six have been together as a crew since June of this year, and after becoming combat ready they were upgraded to senior crew status on August 10th according to Lt/Col John R. Spalding, Jr. commander of the 337th Bombardment Squadron. Lt/Col Spalding also said "They were selected because they were a young, aggressive crew, and were upgraded very fast."

Lt/Col Spalding said that during the last Operation Readiness Inspection (ORI) test September 8th through 12th the crew "flew a successful mission under most difficult circumstances."

During the mission the B-52E flown by the crew developed landing gear, radar, and electronic countermeasures (ECM) troubles. The left aft landing gear would not retract after takeoff and remained down during the mission. The crew made a decent in altitude in order to accomplish in-flight refueling. Captain Shuler, pilot, recently received an evaluation of "highly qualified" during his annual instrument flight check and stand board evaluation. (Post comments of A. C. Mizner. With the drag of the landing gear in the down position, the aircraft became thrust limited as the aircraft gross weight increased to achieve the ORI off load fuel requirements and aircraft gross weight requirement of 420,000 pounds. I used the toboggan procedure to complete the in-flight refueling. I learned the toboggan procedure in the B-47E while in-flight refueling with KB-29 and KB-50 during reflex duty in Alaska in the late 1950s.)

The average age of the crew is 28, making them one of the youngest crews in the 337th Bombardment Squadron's.

(End of newspaper article.)

Crew E-15 was again chosen as the 337th Bombardment Squadron, 96th Strategic Aerospace Wing's Crew of the Quarter for April, May, and June 1965. Crew E-15 was chosen due to outstanding bombing accomplishments during this period. Also During that period Buck Shuler would leave the crew to start his upgrade to B-52E Aircraft Commander. Captain Gerald W. "Gery" Putnam would join the crew as pilot.

As a result of the accomplishments of DYS E-15 crew, in July 1965 DYS E-15 was selected to attend the Top Team Briefing at 2nd Air Force. The crew was flown to Barksdale AFB for three days and two nights stay. During the stay, crew E-15 along with other selected B-52 crews, was given tours of the 2nd Air Force Command and Control Center and received a classified briefing on the current Emergency War Plan. Also, each crewmember received career counseling. The event was topped off with Mess Dress dinner in the Officers Club.

For many years Dyess AFB, Texas had been a B-47E base, but in the early 1960s the B-47s started a phase out to be replaced by B-52s. I was a member of one of four crews to arrive at Dyess AFB in September of 1963 as senior copilot on crew S-01. The crews came intact from Barksdale AFB, Louisiana; Columbus AFB, Mississippi; Mather AFB, California; and Wright-Patterson AFB, Ohio.

After the B-47Es departed Dyess for storage in the Air Force Aerospace Maintenance and Regeneration Center (AMARC) located on Davis Monthan Air Force Base, Arizona, the runway keel was reconstructed to support the increased gross weights of the B-52.

Due to the fact the newly assigned B-52 crews had not flown the B-52E model and while the Dyess runway reconstruction was underway, the crews would rotate via Base Flight C-47, using the taxi way for takeoff, to Roswell AFB, New Mexico for B-52E model difference training. After the training was completed, the crews were given a stand board evaluation and certified combat ready in the B-52E. The crews then returned to Dyess for Emergency Warfare Operations (EWO) training and Wing Staff certification. Following this training the crews were certified by Col Harold A. Radetsky, the 96th Aerospace Wing Commander. The Wing then picked up a SAC EWO assignment and the crews assumed nuclear ground alert in January 1964 at Roswell AFB, New Mexico. The crews remained at Roswell until the Dyess runway construction was completed. In March 1964, the crews started pulling nuclear ground alert and began flying training flights at Dyess. By then, others crew members were arriving forming new crews. As I was on the aircraft commander upgrading list from my prior assignment, a new crew was formed with me as aircraft commander.

I observed over my many years in a combat crew force, those crew members that work together to achieve higher goals become very successful in their military careers, and upon retirement are also very successful in civilian life.

The following are comments that attempt to bring the 1964/5 articles up to date in 2008:

The 337th Squadron Commander Lt/Col John R. "Bert" Spalding, Jr. and A. C. Mizner would again meet in Southeast Asia in 1970. He was then a Brigadier General in 7/13 Air Force. He also was credited with destroying a MiG-15 in aerial combat during the Korean War and was a former F-84F Thunderbird pilot. After 32 years of active duty, Bert retired as a Major General on 1 January 1980.

The 96th Strategic Aerospace Wing Commander Col. Harold A. Radetsky and A. C. Mizner would meet again after we retired. In 1979, when I hired on with General Dynamics (now Lockheed Martin Aeronautics) in Fort Worth, Texas, Harold worked in the Business Management section of my department. In March of 2008, I called and talked with Harold. He told me he is now 90 years old and still runs two to three miles each morning. His voice is real strong and he appeared in good spirits.

Ellie G. "Buck" Shuler, Jr., a 1959 graduate of The Citadel in Charleston, South Carolina, would leave the crew in early 1965 to upgrade to aircraft commander and command a new crew. Upon leaving the crew force, Buck received a masters degree in management from Rensselaer Polytechnic Institute in Troy, New York and then trained as an aircraft commander in the F-4C. He flew 107 combat missions with the 558th Tactical Fighter Squadron at Cam Ranh Bay Air Base, South Vietnam, and participated in the 558th Tactical Fighter Squadron's operational deployment to South Korea following the USS Pueblo Crisis where he flew 15 combat support missions and 57 training missions along the Korean demilitarized zone. Buck would return to SAC later, serving as commander of two B-52 Wings, two Air Divisions, SAC Director of Operations and Commander of 8th Air Force. His first B-52 flight occurred 12 December 1960 and his final flight on 28 April 1991. After retirement as a Lieutenant General, Buck served as Chairman of the Board and CEO of The Mighty Eighth Air Force Museum located near Savannah, GA.

Gerald W. "Gery" Putnam, a 1957 US Naval Academy graduate at Annapolis and former B-47 pilot would leave the crew in early 1967 to upgrade to B-52E aircraft commander and command a new crew. Gery also became an instructor pilot and assigned to the wing Stan/Eval. Gery did an Arc Light tour from January to June 1969. Following that, he attended the Armed Forces Staff College and then went to Vietnam in the C-7 Caribou. Upon his return in 1971, Gery was assigned to 15th AF HQ in training and left as DOTTA in 1974 to attend the Naval War College. Gery was selected to be part of the faculty of the Armed Forces Staff College from 1975 until retirement in 1978 as a Lt/Col. Gery lives in Virginia Beach, Virginia and works as a lifeguard and aquatic instructor at his local Virginia Beach recreation center three days a week.

Earl J. Farney, a 1958 US Naval Academy graduate at Annapolis and former B-47 radar navigator stayed in the B-52 for some time before becoming a staff officer and then Air Base Wing Commander at Vandenberg AFB, California. Earl retired in 1985 as a Colonel and teaches Junior ROTC and Aerospace Science in Del Campo HS in Fair Oaks, California.

Norman D. Elder, after his military commitment of five years, separated from the USAF and became a very successful businessman and now lives in San Antonio, Texas.

Barry Gomborov graduated from the Reserve Officer Training Corps program with a Bachelor of Science in degree in Textiles from the College of Textiles, North Carolina State University in 1959 and commissioned as a second lieutenant and entered active duty in July 1959. After the normal USAF training for newly commissioned officers he was sent to Mather AFB, California for navigator training and upgrade training to electronics warfare officer. Barry then was assigned to Castle AFB for B-52 training. Upon completion of B-52 training, he was assigned to Dyess AFB, Texas on B-52 crew E-15. After his military commitment of five years, he separated from the

USAF and received a degree in accounting and worked as an accountant earning his CPA license. He then set up his own accounting firm and after a few years he went to law school. Barry passed the Bar Examination (first time) in November of 1976 and died 28 December 1976 of pancreatic cancer at age 39 years and three days. Barry and wife Kathy had two children a boy and a girl. Barry is buried in a Jewish cemetery with his family in Baltimore, Maryland.

James T. "Jim" Starr, a full blooded Cherokee Indian from Oklahoma, continued in the B-52 gunnery field and retired as an E-8. Jim was Squadron Gunner of the 20th Squadron, 7th Bomb Wing in 1970 and 1971. During this period of time, Jim's wife Mary came down with cancer. In order to take care of her, Jim transferred to Dyess AFB where there was family assistance as care-givers. After Mary died, Jim retired and within a few months he died of a broken heart. Jim and Mary are buried in Westbrook Texas or about 75 miles West of Dyess AFB. Jim and Mary had no children.

In 1966, the B-52 force was upgrading the old WWII style round dial primary flight instruments to the new Flight Director System. I was selected to attend the Instructor Pilot Instrument School (IPIS) at Randolph AFB, Texas. I flew the T-39 which had the Flight Director System on the left instrument panel and the old round dials on the right instrument panel. I graduated in January 1967 and returned to Dyess AFB to teach the flight crews the proper use of the Flight Director System. In addition, I was certified to teach the annual instrument school and administer the annual instrument examination. In addition, I was flying crew sorties, pulling Supervisor of Flight (SOF), teaching the Flight director System to the crews on alert and administering the annual instrument examination. I was also the assistant Flight Safety Officer and had to keep all the records updated and conduct the required safety meetings while the flight crews were on alert. I also graduated from the B-52F Central Flight Instructor Course at Castle AFB. In addition, I graduated from the SAC academic instructor course. In September 1968, I was assigned to Barksdale AFB, Louisiana 1st CEVG and spent two of the next three years deployed to SEA on special assignment. I was able to fly as an advisor on several different types of aircraft during this period. In September 1971, I was assigned to 20th Bomb Squadron Carswell AFB, TX flying B-52Ds. On 1 January 1972, CAR E-57 helped form the new 9th Bomb Squadron. In February 1972, the 9th and 20th Bomb Squadron personal were taken off ground nuclear alert and told to go home and pack for deployment to SEA. I was deployed to SEA and finally returned to Carswell AFB in June of 1973 to be an instructor pilot (IP) in the RTU and 9th Bomb Squadron IP. I would fly three times per week plus pulling SOF each week

Since I entered the USAF under the Aviation Cadet Program right out of High School, I lacked the college degree needed for the new Air Force. I was able to accomplish 43 semester hours of college during my military career. I also knew I needed a college degree to compete in civilian life. My wife said she would get her nursing degree to help support the family and then I could retire. I retired 31 December 1976 and on 16 January 1977, using the GI bill, started college at Texas A & M. I carried a very heavy student load and graduated in December 1978 with a degree in Industrial Technology. In January 1979, I was accepted into the masters program. At the same time I applied for employment at General Dynamics, Fort Worth, TX. In May 1979, I had my interview and was offered a position starting the following week.

In 1996, the company paid for me to go back to school an get a Social Science degree to better understand our new employees and how we might motivate them into being career employees.

Now 29 years latter I am an F-16 Aerospace Staff Engineer in support of the F-16 for the USAF and all foreign customers. During these years I have traveled to many USAF bases and several foreign countries. My current plans are to work another eight years to get my grandchildren through college and retire at age 82.

Without the Christian ways taught to me by my parents, the discipline of the Aviation Cadet program, and the unusually strict rules of SAC, plus many combat missions, my civilian career might not been so successful as it has been.

B-52D Nav station.

The Consequences of
One Low-Level Abort
Dale Fink

This story is for those of you who have endured circumstances similar to mine. As many of you know, most of these kinds of stories usually go untold and the facts are just "swept under the rug." That's what makes this series of books important. If you know what I'm talking about, this one's for you. If not, this story is to help prevent similar situations for those who follow in our steps as Crewdogs.

I need to say that the references to distances and specific procedures should only be viewed as relative references in the following text, as I forgot the actual numbers, rules, and detailed procedures as the door hit my butt on my way out of the Air Force.

It was in the mid-1970s, and my Carswell AFB, Texas crew was flying a low-level profile training sortie at La Junta, Colorado in the fall. As we approached some mountainous terrain, the weather began to worsen, and quite quickly we found ourselves with severely limited visibility.

On one leg of the low-level route we were supposed to fly straight at a mountain peak that was significantly higher than our flight altitude.

My job as Navigator was to call out the distance to the peak as measured on my navigation radar. Our SAC procedures directed us to turn away from the peak when we were two miles from it, and the pilots were supposed to maintain visual contact with the peak once we were inside five miles and echo (confirm) the distances as I called them out. That was a mandatory SAC flight safety procedure, according to "the book."

Just as we hit 10 miles from the mountain peak, the copilot announced that he had no visual contact whatsoever outside the plane. The pilot claimed he could see "limited" landmarks. The copilot reminded him that the rules were that BOTH of them had to have visual contact once we crossed the five-miles-to-go marker.

Our pilot was known to always be right on the edge of dangerous every time we flew and had a bad reputation for "bending" flight rules. No one could ever straighten him out unless they wanted to incur the wrath of the squadron commander, who was our pilot's self appointed mentor and drinking buddy. So I was a bit surprised when our copilot "reminded" our pilot on the safety issue while in flight. It could only mean that he was REALLY scared and wanted to let the other crew members know it.

We continued flying low level in "the soup" at five to six miles a minute, so keep in mind that the following conversation occurred in about a minute or so. As we reached the five-mile mark to the peak, I called out "Five miles to go." and added "Confirm visual contact." The copilot replied "Negative visual contact," then the pilot replied, "Hang on, I think I see something." I almost immediately came back on the intercom saying, "Four miles to go; negative visual contact; abort route now." It was my responsibility to request that the pilot abort the route, and it was the pilot's responsibility to actually initiate the call. The pilot immediately replied, "Negative on the abort; I think it's clearing up ahead." I switched over to guard frequency on my communications radio, and as we passed the three-mile marker, I announced over the guard channel "(Our call sign) aborting La Junta low level route at this time. Abort! Abort! Abort!" Later, I found out that the copilot initiated a climbing turn as soon as I broadcast the low level route abort because the pilot was too busy screaming every profanity known to man at me at that time.

28

The pilot vented his "opinion" all the way home; however, as soon as the wheels touched down, he shut up. I guess that was because he knew that "officially" he couldn't object to my following procedures, and he didn't want to have to admit he had tried to "bend" the rules, again. Of course, as I opened the hatch, I saw our squadron commander there ready to "debrief" our pilot.

After our pilot and the squadron commander had time to get their stories in sync, I had a private meeting with the commander where I was told (off the record, of course) that even though I had followed procedure correctly, I had displayed a "wanton disregard for the unwritten rules of the 'team player' mentality" and had not tried my best to fulfill our mission requirements. I guess saving a multi-million dollar aircraft and her crew from flying into "cumulous-granitus" wasn't the right "attitude." He then explained that I needed to be given some time off from flying - time to think about my attitude and so forth. The very next week I was assigned to alert duty to replace a "sick" navigator who miraculously recovered overnight and was actually flying by the second day. I then began a 70-day stint of replacing "sick" navigators on alert.

When I finished my "two-month-plus" alert cycle, I was transferred to a crew that had an "attitude" just like mine! I never enjoyed flying in the BUFF so much as I did when I was assigned to this "attitude" crew. My new crew flew with safety as our first priority, and everything after that was just gravy. None of us ever received a "good deal," however, from that point on. Most of us, in fact, left the service when our commitment was up. Those that stayed in never advanced very far in rank, but, by God, every time we took off you could count on us coming back safe and sound!

Bill Jackson

Those Early BUFF Days
Bill Jackson

Very little has been published about the early days of the B-52 program. Since I was involved during those days, allow me to relate some of what I remember about them.

Basically the B-36 wings provided the first crews to begin CCTS at Castle AFB, California, and as the training progressed additional crew members were provided from the B-47 Wings. The B-36 pilots were all experienced high timers in everything required of pilots except air refueling and the ability to think at eight or nine miles per minute. Their aerial refueling experience consisted of a single orientation flight in a B-47 with refueling from a KC-97.

I first became aware of the buildup of the B-52 fleet when several combat-ready crews departed for Castle early in the summer of 1957. My crew was given "E" status at this time with a new copilot and orders for Castle in October. A prerequisite was we had to pass the Combat Evaluation Group evaluation which applied to all E crews. We passed and when I went to the personnel shop to get my copies of transfer orders, I discovered that we had volunteered for the new B-52 program. The copilot was on an indefinite status and I had the 3,000 total flying hours required. When I pointed out that I was a year too old, I was notified that SAC had waived that restriction!

Ground training began in October 1957 with flight training following in December. Since December was usually a month of fog which plagued the Merced area, my class was instructed to go home and not to come back until the New Year. During that time period the existing B- 52 fleet was suffering from leaky Mormon clamps in the fuel system. When we did finally get on the flight line, the maximum fuel load was enough for only a four or five hour mission since only the main tanks could be used. The instructor crew that we were assigned was one of the three crews that had flown in the around-the-world record flight of a year or so before. On the fourth mission the IP found a KC-135 tanker in the local area, and that was when I found out that one did not refuel at 35,000 feet altitude!

In February 1958 we transferred to Fairchild AFB, Washington, after only five training missions. We arrived just as the wing became combat ready. All their old B-36 crews had regained their spot promotions, so we were relegated to a status of merely being tolerated. The tanker squadron was getting its birds so air refueling had not yet been scheduled. Later the procedure evolved to where a B-52 would take off and in the blind use the UHF radio to find a tanker in the area that could give a few minutes of contact time. Any C-135 would do including Boeing Company crews. About the time of my first standboard check tankers were available on a more or less routine schedule so all flights could have a shot at refueling.

We developed schedules and procedures as we went along. At the beginning of my training crew assembly time was three hours prior to takeoff. Fairchild was the aircraft engine depot in the states for the entire Pacific theater during WW2 so there were more than enough large buildings on base. The main hanger could hold more than four BUFFs. I never counted them but I would estimate there were at least ten or more enormous warehouses. One contained the base gym and a huge swimming pool. One was used to store boats and travel trailers belonging to base personnel. Imagine a heated storage area with a sprinkler system to protect all that in the savage winter months. The crew assembly area was in a building that had lots of room for crew lockers, a parachute storage area and a helmet and oxygen mask shop where those pieces of gear were stored and cleaned,. There was also a secure area for side arm storage and enough empty floor space for several crews to lay out their gear for inspection prior to boarding the crew bus for transport to the flight line.

The only big change in preflight procedures between the B-47 and B-52 was the requirement for the flap well and the engine intake areas to be inspected by the aircraft commander. This required the use of a work stand with the pilot onboard to be pushed from the wing root area towards the wing tip from one outboard engine to the other tip and then up the flap well to the other wing root. This was quite an accomplishment especially if the ramps were covered with snow and ice.

In the past all sorts of debris such as tools, lunch pails, rotten fruit and similar objects left by Boeing factory workers had been found in the flap well and engine intake areas.. On one occasion a fuel cell was removed and a lunch pail was discovered inside. This practice soon came to a screeching halt.

Let me explain my remarks made earlier about the old timers and their inability to think at a speed of eight miles per minute. When a wing mission was scheduled, a tanker would takeoff first followed one minute later by its scheduled receiver. This was repeated by the follow-on aircraft. The lead bomber was always flown by the standboard chief pilot and crew.

During one refueling a bomber was too close to the tanker to get its offload at the scheduled time. The bomber pilot overshot the tanker. In order to get back in the proper position he proceeded to make a 360° turn to the left. Needless to say, this was an absolute no- no. At debriefing in his defense he stated that no one ever had told him that was not the thing to do!

You can see how things were during the early days of the BUFF. I am sure today's crews would consider this as being ancient history. Since the B-52 is programmed to remain in the USAF inventory until 2040, the procedures we used in Vietnam and subsequent skirmishes will eventually be considered ancient also - like those used in my first flight in 1958.

A Dream Realized
John R. Cate

The following story is dedicated to two of SAC's many flight instructors. Both men were highly motivated and dedicated to a cause in which they truly believed. Their daily sustained efforts put the "iron" in the Strategic Air Command's gloved fist and helped to make her the most feared and respected air command the world has ever known. Lest we forget.

Finally, the telephone call I'd waited for over three months had come. The commandant's secretary brought the message to my classroom herself. It was January, 1980 and the call had come from CMSgt Timlake, the B-52 Gunnery Program Manager at HQ AFMPC, Randolph AFB, Texas. The message directed me to contact the Chief as soon as possible to confirm the dates for my class III flight physical and altitude chamber! I reread the message twice, just to make sure. The next classroom break I called Chief Timlake and was told my flight physical was scheduled for that Wednesday at the base clinic and my altitude chamber was scheduled for the following Monday at Peterson AFB, Colorado. Talk about making things happen. I'd had only known of one Chief Master Sergeant in my brief Air Force career and to be certain Chief Timlake was not like him at all. At the time I was an ATC classroom instructor stationed at Lowry AFB, in Denver, Colorado teaching the TAC F-4 Weapons Systems Course at the "Black Shack". I'd been in the Air Force for seven years and that was only my second assignment. My first had been as an F-4 weapons release mechanic assigned to the 81st TFW, RAF in Bentwaters, England.

In the summer of 1979, myself along with two fellow instructors (Jim Rogers and John Gault); had applied for retraining as B-52 Aerial Gunners. My dad had been a B-24 top turret gunner/flight engineer flying out of New Guinea and Port Moresby, Australia during World War II. He flew combat for over two years and as with many other WWII veterans, never talked about his experiences. Like so many boys of my generation, I wanted to be like my dad and "follow in his footsteps". I'd taken my first airplane ride in grade school, a Piper Cub piloted by one of my dad's friends, R.C. Fox. He was a World War II veteran who had trained as a C-46 copilot to fly the "Hump" out of the China-Burma Theater of Operations. The ride cost $5.00 and the money was used to install a public address system in my elementary school. I saved my allowance for two months and did extra chores for my mom around the house to raise the money. During the 15 minute flight, I saw my house from the air and from that point on I was hooked. I ate, breathed, and slept, all things airplane and the Air Force.

At the time I made my request for retraining, the personal specialist explained that my career field was "short" and the gunnery career field was "balanced" in my year group. He also reminded me I was on a controlled five year tour as an instructor and I could only request a one year wavier. I had served only three years so it left me one year short. He said he had been directed to send my request forward, but he was sure it wouldn't be approved. I called CMSgt Timlake and told him what I had been advised. The Chief asked me not to worry; said he would take care of everything. He also wanted to know where I would like to be stationed. I told him Robins AFB, Georgia as it was the closest SAC base to my hometown in Tennessee. I still hadn't put two and two together and naturally didn't hold out much hope. I told myself if I couldn't fly, Denver, Colorado wasn't too bad of a place to be stationed.

Time passed slowly. During the second week of February 1980, another call from CMSgt Timlake finally came. He began by saying that my TDY orders to attend B-52 Gunnery Training at Castle AFB, California were waiting for me at CBPO. Then the Chief told me he needed me in place at Castle AFB no later than Sunday to begin my gunnery training Monday morning! At last, it clicked! With the assistance and support of CMSgt Timlake I had managed to take the first of what would be many steps in becoming a B-52 Gunner.

Sunday afternoon I checked into billeting and found there was a message waiting for me from SSgt Pat Fagan. The message told me to report to the 4017th Combat Crew Training Squadron (CCTS) or "schoolhouse," as it was more commonly referred to, at 0800 the next morning. After unpacking and settling into my room I decided I better find where the 4017th CCTS building was located. I found the schoolhouse in short order along with the Base Exchange, commissary and the NCO club. On the way back to my room I stumbled across the flightline. As I stood in awe at the fence and looked at the two rows of B-52s, the most beautiful aircraft I had ever seen. There was a row of G models and a row of H models. For a brief moment, I wondered what the future held for me. Although I couldn't be certain of anything at that point, I knew I would make it.

The next morning I arrived at the schoolhouse at 0745 to begin the academic phase of my B-52 gunnery training, the second phase would be flight training. I soon found my fellow classmates. They were easy to spot. There were six of us, including myself and we all had an uncertain look on our faces and, of course, no enlisted aircrew wings on our chests. Two of my classmates were SSgts like me, Kelly Wolfe and Don Miller. There were also three airman, recent graduates of basic training. Four of us were destined to become G-Model gunners, the other two H-Model gunners. At 0800, we met our academics instructor, SSgt Pat Fagan. He introduced himself and showed us to our classroom. I liked Pat right away; he was open, approachable and seemed glad to meet us. He told us a little about himself and what the next eight weeks had in store for us. Pat told us the academic phase of the course would be tough, but if we applied ourselves, we would pass. He said this meant studying every night and on weekends too.

He also told us something I had been hoping to hear, each of us would be taking a familiarization flight after three or four weeks of training, just to make sure we really wanted to be a B-52 gunner and to get our flight pay started. Pat further explained that as enlisted flyers we needed four flight hours a month to qualify for flight pay. Next, we were issued a Dash 1 flight manual that covered the B-52 in general and also contained the emergency procedures or EP section, a Dash 5 flight manual that covered the operation and theory of the Fire Control

System or FCS along with a section that was used to trouble shoot any FCS malfunctions. We were also issued an In-Flight Checklist that was used to operate the gun system correctly and safely in flight along with a separate emergency procedures section. The In-Flight Checklist was an abbreviated version of the Dash 1 and Dash 5 and contained procedures for doing a FCS modes check and appropriate emergency procedures, to name a few. In addition to our flight manuals we were given a set of binders and spent the rest of the morning putting our flight manuals and checklists together. After lunch we did the first of what would be many page counts on our Dash 1, Dash 5, and Checklists.

Just prior to the end of our first day, Pat opened the floor for discussions. Of course, all of us had a dozen questions, pretty much all the same. The two main topics of discussion were; "What's it take to be a gunner?" and "How much is our flight pay?" Pat's answers were: "Hard work, desire, and determination." and "Not enough." Pat told us to familiarize ourselves with the Dash 1 that evening and bring our Dash 5 and Checklists to class the next morning.

The next day class began at 0800 sharp. Pat immediately separated the class and the two H-Model gunners left with another instructor. We would see our fellow classmates from time to time, but for then, it was time to begin our G-Model training in earnest. I looked around our classroom and for the first time I noticed a chalk board mounted on a pedestal with white lines in the middle, the outline of a box with handles on the left side and on the right side, an outline of what looked like a radar dish. During the next eight weeks that chalk board would become my best friend. We spent the morning going over our in-flight checklist, covering the "after engine start" and "modes check" sections. After lunch the four G-Model gunners, myself included were taken to the T-1 building. The T-1 Trainer was a G-Model FCS mockup made to look like the actual gunner position on the aircraft. The T-1 Trainer had many useful purposes, for instant it could be used to introduce FCS malfunctions and was a very useful training device until the Weapons System Trainer or WST came on line in the mid-1980s. A T-1 Operator and Gunner were assigned to the T-1. SSgt Johnny Mize was the instructor for our first T-1 training session. Johnny was another good instructor and laid the foundation for my understanding of the fine art of tuning the G-Model scope for optimum presentation. SSgt Mize spent that afternoon teaching us the basics of checklist use and how to "power up" the FCS. I found out that the G-

Model gun system was called the ASG-15 and had 23 modes of operation, 24 if you counted "Off." It was a long afternoon.

And so it went, day after day: classroom work, studying the Dash 5 and Checklist, T-1 trainer sessions, and homework every night. Friday marked the end of our first week of training and our first block test. It was a bear, but I surprised myself and did better than I thought. In week two, Pat arranged for us to have a tour of a G-Model. I had flown commercial a grand total of four times at that point in my life. The B-52G was different from any aircraft I had ever seen. She was an aircraft built with a single purpose in mind - she was built for war. She was designed for speed not comfort. My dad, when he spoke about a B-24, and he rarely did, called her a "Thing of Beauty". The B-52 was destined to become my "Thing of Beauty". Friday, week two and another block test. Again, I pulled down a good score. But, I had studied on the average of four hours every night and all day, Saturday and Sunday. I allowed myself Sunday afternoon off.

During week three we were issued a basic issue of flight gear, two flight suits, boots, flight gloves and a flight jacket. We were also issued a flight helmet which we in turn, gave to life support for fitting. During week three we also had our fingers and feet printed, in the event of an aircraft accident. Friday of week four and our first classmate had his familiarization flight, an eight-hour sortie. You guessed it, airsick from level off through full stop. Don, Kelly and I visited our fellow classmate on Saturday and it was plain to see that he had a rough time. He talked about self eliminating. After meeting with Pat Monday morning, he decided to stay. Week four, Kelly was next, than Don. Both flew on the same day and had good flights. I got my chance at the beginning of week five. Over the years, I've forgotten the name of my flight instructor, but I can still see his face and hear his voice. He came by my quarters Sunday night and said he would pick me up Monday morning at 0730 and that we would be mission planning at the 328th Bomb Squadron morning briefing at 0800.

Mission planning went smooth, but I was surprised by how long it took - over eight hours. After mission planning, I stopped by the life support section and was fitted with a "loaner" flight helmet and oxygen mask. Showtime was 0445 the next morning at base ops with a 0600 takeoff for 6.0 hours in duration. I was flying with a crew on their solo mission. My first flight is a blur now, but that day I found out what a flight instructor is. He was with me through every phase of the flight, from preflight to postflight, keeping me in the checklist, challenging me, encouraging me, keeping me safe in a totally alien environment. I would attend instructor upgrade some six years later and his was the first technique I thought of. I was back on the ground and in class after lunch; sorry it was over and anxious for the next flight, knowing I had found my calling.

The next three weeks passed quickly at the schoolhouse with a routine established. Block tests every Friday, T-1 trainers and sessions with the chalkboard. The chalk board was used to "diagram" the different modes of operation of the ASG15, what components were and weren't used in a particular mode of operation and helped us determine what would be the "optimum" mode with a certain malfunctioning component. For instance, Pat would ask one of us to step up to the chalkboard and diagram a certain mode of operation, such as Radar Track/Radar Range or Manual Track/Manual Range. He would take a seat with the rest of the class and you were on your own, diagramming from memory the mode in question. We did these exercises for obvious reasons, in combat the correct split second decision could mean the difference between defending your aircraft or not.

Friday, week eight, graduation day. Our final block test. All six of us made it through. In addition to a graduation certificate each of us were awarded a pair of basic enlisted aircrew wings in a simple, but memorable ceremony. The 4017th Squadron Commander, SMSgt Norm Lake the NCOIC of gunnery training at the schoolhouse, and the 92nd BW Wing Gunner, CMSgt Jerre Albright were in attendance. SMSgt Lake presented our graduation certificates and the squadron commander gave each of us a pair of wings. I still have that first pair. CMSgt Albright than gave a short speech on what we had accomplished, what lay ahead and most importantly, what he expected of us. Before releasing our class for lunch, Pat told us we would be meeting our individual flight instructors and the crew we were assigned to for the flight phase of our training. I thought I was ready; I was in for an abrupt awakening.

After lunch, our class along with Pat met at the Officer's Club. There we were introduced to the crew to which we were assigned for the flight phase of our training. The members of my crew were: Pilot-Capt Russ Sypolt, who was upgrading to Aircraft Commander or AC, from Robins AFB; Copilot - 1Lt Darrell Stinger, recently graduated from undergraduate pilot training or UPT at Columbus AFB with a follow-on assignment to Robins AFB; Radar Navigator - Capt. Terry Phillips an upgrade from Navigator to RN from Robins AFB; Navigator - 1Lt. Gerry Valentini, who recently graduated from undergraduate navigator training or UNT at Mather AFB, with a follow-on assignment to Robins AFB; and Electronic Warfare Officer or EW - Capt. Vick Dilda, undergoing re-qualification training with a follow-on assignment to Robins AFB. Over a few beers, we talked and got to know each other, and I felt I had been assigned to a good crew. I was right.

That afternoon I also met my flight instructor, MSgt Bill White. Bill had been an AC-130 Gunship Gunner during the Vietnam War. As with Pat, I liked Bill from the beginning. He had an air of confidence about him, not at all cocky; unassuming, but sure of himself. Bill was approachable and easy to talk to. Over the next 10 weeks he would finish what Pat had begun. Yes, I had a working knowledge of the ASG-15 gun system, but nothing more. It would be up to Bill to teach me how to be a crewmember, how to bond with my EW to become the defensive team of our crew and, in short, how to fly in the B-52G safely and effectively. It would be up to me to learn. Keep in mind that as a gunner going through initial gunnery training I was the only member of his crew with no prior flight experience. The AC and RN both had met flight hours and years of experience requirements to be eligible for upgrade; the CP had recently graduated from UPT; and the Nav and EW from UNT.

Consider the following list of inflight duties of a gunner: the gunner had to become comfortable with performing in a flight environment to include low level flight, monitor and report all passing traffic, effectively and safely operate the Fire Control System (FCS) both inflight and on the ground, and configure the FCS for its proper mode of operation to effectively defend the aircraft. He also had to coordinate and perform inflight system checks with fellow crewmembers, defend the aircraft against any and all threats to include fighter attacks, simultaneously monitor three radios (2 UHF & 1 HF) and the Guard Channel, monitor and respond to all interphone calls. Duties also included the need to accomplish the correct checklist during the corresponding phase of flight, monitor the ALR-46 Warning Receiver, monitor and operate the SATCOM or Satellite Communications System, find a way to integrate himself and get along with his five officer crewmembers and occasionally overcome a bout of airsickness!

Of course Bill knew the uphill climb I faced and he quickly became the steady voice in my ear that challenged me, cajoled me and questioned me every time we flew. He taught me how to mission plan, preflight, postflight, how to word a FCS write-up for the Aircraft Form 781, how to debrief the FCS technician at maintenance debrief. If Pat had taught me to crawl, Bill taught me to walk. Before we left that day, Bill gave me homework for the weekend. "Familiarize yourself with the Dash 1, especially the Emergency Procedures Section" or Section III as it was commonly referred to and be ready for an EP test on Monday.

Monday morning, I took my first of what would eventually be at least 100 EP tests during my B-52 career. I passed, barely, by one question. When Bill had graded my test, he simply looked at me and said, "Not good enough. I expect you to make nothing less than 100%, every time." Then he requested my Dash 1 and opened to page one of Section III. There in bold print, about halfway down the page were these words - STOP, THINK, COLLECT YOUR WITS. He went on to explain that the emergency procedures were based on other crewmembers experiences. Knowing them and understanding them, Bill said, would improve my odds surviving a ground or inflight emergency. He had my attention. We spent the rest of that morning reviewing emergency procedures.

After our lunch break Bill spent the remainder of the day explaining the training syllabus. The syllabus told me that my crew and I would fly at least 12 missions to accomplish all the training requirements and four of these would be night sorties. If weather prevented us from accomplishing a certain training requirement such as flying a particular low level route or an aircraft malfunction prevented us from accomplishing a required training requirement such as an air refueling, or if one of my crewmembers or I failed to meet a certain training requirement, then we would be flying additional missions. Bill pointed out that if all went according to plan, I would have my checkride on sortie 12 and my solo flight on sortie 13. I also learned that I would be on a three-day training cycle. Day one - mission plan the sortie, day-two - fly the sortie, day-three - debrief and critique the sortie. During our night flight rotation, the training cycle increased from three to four days with day-one - mission plan the sortie, day-two - fly the sortie, day-three - crew rest, day-four - debrief and critique the sortie. Time frame to complete the training syllabus and fly the required 13 sorties was normally eight to 10 weeks.

On Wednesday, Bill and I mission planned for my first training sortie. The flight went smooth but to Bill's disappointment the FCS was Code 1. I thought I had done well and I had. The only trouble was I stayed about thirty minutes "behind the aircraft" all day. In other words, all my actions were performed late because I wasn't thinking fast enough. I knew I could change that area of my performance and I did. Bill scheduled extra T-1 trainer sessions and gave me homework every night. I always had the checklist with me. Anytime we had a few minutes, we would "table fly" a portion of a mission, with Bill giving different scenarios, different malfunctions, and asking one "what if " question after another.

Unknown to me at the time, each flight was structured to build on the one before it. As a result I got a little better each time. By the time my crew began our night flight rotation, I had hit my stride. No longer was I "behind the aircraft", I was actually staying "even with the aircraft" and on some occasions, actually "ahead of the aircraft". Bill's patience and persistent were starting to pay off. Sortie nine was our first day flight after completing night flight rotation and we were scheduled for a Fighter Intercept Exercise or FIE with the California Air National Guard flying F-106's out of Fresno. I did okay, had my first experience with airsickness but the EW and I successfully defended our aircraft.

I had two sorties left, 10 and 11, before my first experience with Stan Eval. Bill never let up and I'm thankful for that. On Tuesday, I took my two required Stan Eval tests - an open book general knowledge test consisting of 100 questions and a closed book emergency procedures test consisting of 25 questions and four bold print procedures such as Rapid Decompression or Controlled Bailout. I was successful with both. It was quite a difference from that first EP test I had taken just nine weeks earlier. I mission planned for my first checkride on Friday with the flight scheduled for the following Monday. It would be a crew effort since we were all taking our check rides together. That twelfth fight went smoothly. We delayed takeoff due to our mated tanker having maintenance problems. I managed to "stay ahead of the aircraft" all day. Bill met me at Base Ops after the flight with the number one question any flight instructor asks his student - "How do you think you did?" I told him I was sure I had done okay, and had no problems with FCS operations or crew coordination and I had configured the FCS for its optimum mode of operation. I did feel I may have had a problem with SATCOM Operations though. I was required to transmit and receive three messages during the flight. I had done so but had somehow used the wrong message format for one of the messages. I knew mission planning had gone smooth and there were no problems or mistakes with my paperwork.

My checkride debrief was scheduled for 0900 on Tuesday morning in the Wing Gunner's Office. Bill and I arrived at 0845. Those next 15 minutes were the longest 15 minutes of my life. My evaluator was there and he set across the room from us, never saying a word. Precisely at 0900, CMSgt Albright asked that we come in his office. We all set around his conference table, Bill and I on one side, my evaluator and the Chief on the other. The debrief was lengthy and detailed. I had no major areas of concern but I had taken a write-up for SATCOM Operations. Otherwise the evaluator said, "I liked what I saw and overall it was a good flight".

Bill offered his congratulations and we shook hands. The Chief said my name and as I looked toward him, he too extended his hand. Instinctively I reached across the table to shake his hand, he said, "Congratulations, John" and placed something in my hand. I looked down and discovered I had been given a gunner coin. He also gave me a copy of the Rules of the Gunner Coin (Bean). Chief Albright strongly suggested I familiarize myself with them as soon as possible. That evening Bill and I met at the NCO Club for dinner and a few beers.

On Friday, our crew flew its solo flight. The aircraft seemed somehow quieter and less crowded without our cadre of flight instructors. My EW wasn't required to fly a solo mission, since he had been previously qualified in the B-52. Instead I flew with a student EW on his second sortie. As he preflighted his seat with his instructor hovering over him, I realized that had been me only 10 short weeks earlier! As he was about to check his oxygen panel, he discovered he didn't have a CRU-60. He admitted to his instructor he had left it on the aircraft after his first sortie. His instructor asked the Aircraft Commander to radio Command Post and request that Life Support bring one out. I told him not to bother and pulled a spare CRU-60 from my helmet bag and gave it to the student EW. I realized MSgt Bill White and SSgt Pat Fagan had taught me something after all.

For me, becoming a B-52 Gunner was truly a dream realized. It gave me a career which I look back on with a sense of pride and accomplishment. It gave me the adventure of flight and allowed me to serve this great nation. Along with my fellow gunners, we were allowed to make a difference. Can any of us ask for more? Most of all, it has given me the comradeship of my fellow gunners that I have enjoyed for nearly 30 years. If given the opportunity, would any of us do it again? Yes, without hesitation. Was it worth the effort and the sacrifice? Absolutely. In our changing world things have gone full circle and one is left to wonder if there will ever be another Cold War? For a moment consider the current political instability in many, if not all, of the countries that made up the former Soviet Union. It is a distinct possibility. One is left to ask, "Did we really win the Cold War? Or did the Soviet Union?"

To my fellow gunners - if there are any mistakes or omissions in this story, my apologies as it was purely unintentional. As the years slip past, the memory sometimes fades. Gentlemen, it has been my honor and privilege to serve with you. C'est La Vie.

(Thanks Rory - Thanks Jerre)

Chapter 1 – Military Career

"The Flame"
Gary Henley

"You have to be tough to fly the heavies," did not just mean being tough in a physical sense. It also meant that you had to be tough mentally as well. It was not always brawn, but sometime brains and stamina that defined a person's strength.

The mid-70s was replete with turmoil for members of the SAC crew force. The combat-hardened crews had returned to the Continental U.S. (CONUS) from their tours in Southeast Asia, and found themselves once again flying nuclear war training missions instead of real-world combat missions. The "Super" bomb wing at Carswell AFB, Texas, found that it was fat on Vietnam wartime experience, but slim on peacetime readiness and out of sync with the non-wartime SAC rhythm. Even though the United States was still at war; it was the war that many fail to mention when recounting the great history of our nation - it was the Cold War. It was the war that Ronald Reagan proclaimed we had won without firing a single shot. It was a war that Sun Tzu would say epitomized the ultimate accomplishment in warfare - one in which your enemy gives up without any battles. The burning question on most people's minds was, "How do you shift gears from wartime to peacetime alert and EWO readiness?" It turned out to be a tense time of shifting gears from high-altitude, multi-ship B-52 formation flying with iron bombs to low-altitude, terrain following, lone-penetrator bomber missions for proficiency in flying "The Big One".

The returning wartime SAC Crewdogs at Carswell were real pros, but seemed to be under more than the unusual scrutiny than some of the other SAC bases. This was partially due to the fact that an Air Division, headed by a one-star general, was also located on the same base. We were considered a "Super" wing because our wing included two heavy bombardment squadrons, one tanker squadron, a Combat Crew Training School (CCTS) for D-model B-52s, and a Consolidated Flight Instructor Course (CFIC).

In the summer of 1975, I was completing my CCTS training to become Emergency War Order (EWO) certified amid the veritable plethora of Crewdogs that had flown more combat time overseas than I had in my entire Air Force career. It was a little intimidating for a new SAC Crewdog (maybe "Crewpup" is more appropriate); however, it was also inspiring to be among those who had fought in Vietnam, flown Linebacker II missions, and some of whom had even returned to the cockpit following the experience of being a POW in the Hanoi Hilton.

This was not your typical pre-Vietnam SAC Crewdog workforce makeup. These guys had medals all over their chests like I had never seen in my life. The task that lay ahead, however, was like trying to convert the "Dirty Dozen" into "1976 SAC Poster Children." This was the time in the history of SAC that the crew force needed to be molded back into the "spit-polished-boots" necessary Cold War fighting force in order to hold the Soviet Union at bay. The way I see it now was that was that the crews at Carswell needed a mindset change in order to transition into a new mode of operation - one more rigid and regimented in accordance with SAC alert-force doctrine and centered on our EWO mission should we, God forbid, be called upon to execute that mission.

Senior leadership at the time seemed to be more ruthless and overbearing than even SAC's original cigar-chomping icon, Gen Curtis LeMay. Wing commanders had so much pressure on them that they were like cats on a hot tin roof, micromanaging everything we did and running scared (for good reason) for fear that they would make a mistake that would end their careers.

Dave Lay, a pilot at Carswell during that period, recalled returning from a flight one day when he was in CCTS upgrade training. It was about 0430 hours, and his crew had just landed from a night flight. Dave was getting off the bus to finish loading bags after the flight and was literally run over an individual trying to get onto the bus. Dave apologized profusely for running into him and suddenly realized it was the Division Commander, Brig Gen Burpee, who was very short and thus the reason Dave did not see him. Gen Burpee said nothing to comfort him (e.g. "Good morning. How was the flight?"). Instead, he yelled, "Were you the one that didn't get your gas?" Dave politely told him he had in fact taken his scheduled on-load with one contact. Gen Burpee said, "Oh!" and stormed off the bus. Dave thought to himself, "What on earth is the Division Commander doing on the flight line at o'-dark-30 hunting for a crew that did not complete refueling?" In

Dave's mind, this should have been a squadron matter, not a division concern. Yet there was the one-star, ready to hang a Crewdog because he did not get his on-load during a night air refueling (A/R).

Dave was so very glad he had gotten his scheduled on load. After that incident, Dave left the air refueling auto pilot off, refueled autopilot off, and did at least half of every future track with the tanker autopilot off to ensure he would never be in a Division/CC one star's cross hairs for not completing A/R. As Dave states, that incident indelibly imprinted a lesson in his mind, so something good did come from this encounter. There is a story in *WWCD II* about how Dave ran the full gamut of A/R on EVERY flight from that time on.

This bizarre behavior from our scared senior leadership left a huge puzzled impression on me as I entered the operational B-52 crew force. At that time, for example, stories about the Wing Commander's behavior had filtered through the squadrons and into the CCTS squadron to the extent that our CCTS instructors were scared to death that one of their students would not do well in their initial EWO certification. Fortunately, I had done well on my CCTS tests, simulator check ride, and flight check, so I didn't think I had a lot to worry about. After a successful completion of CCTS, each new combat-certified Crewdog would be scheduled to meet with the Wing Commander, who would individually review each student's records and "welcome" him into the elite prestigious SAC Crew Force (a.k.a. SAC's "trained killers").

While awaiting my turn in the hallway for this "rite of passage," I realized that my scheduled time for the appointment with the Wing Commander was already past due. I remember hearing the yelling and screaming coming from behind his closed door as he gave the poor Crewdog ahead of me the verbal beating of his life. It seems that poor guy had almost busted his open book EP (emergency procedures) test, and his check ride was satisfactory, but not without significant write-ups. The berating from the Wing Commander was not without a "colorful" stream of admonitions and expletives that seemed to go on for hours (actually, it was only about 30 grueling minutes). That did not make me one bit comfortable, knowing that I was his next "victim." The door opened, and the thoroughly demeaned and red-faced Crewdog departed the Wing Commander's office. At that point, I took a deep breath, swallowed hard one big gulp, and then moved forward following those dreaded words, "The Wing Commander will see you next."

Keep in mind that at this point the Wing Commander (I will keep his name out of this because I don't want to embarrass him), whom we nicknamed "The Flame" (due to his reputation for being a "flamer"), had worked himself up almost into a frenzy due to the previous Crewdog's debrief. I was the next course on his meal of eating Crewdogs alive, and the next few moments would tell the tale of my survival or death. I do remember thinking, "Thanks, Crewdog Number One! Now I have to face The Flame at the peak of his anger. Can this get any worse? This is going to be one of the worst days of my life, I'm sure."

I reported in to the Wing Commander, saluted smartly and awaited his return salute, which was one of the most abbreviated salutes I can recall. "Take a seat!" he barked, and I timidly sat down on the edge of the chair in front of his desk in dire anticipation of what he would do or say. Before the "bloodletting" of the previous Crewdog that I had just witnessed with my ears, I expected something like a pep talk that coaches give their players once they had made the team - something like, "You've been through a lot of training, Lt Henley, and now you're a proud member of the SAC crew force. I expect you to carry on the finest traditions of SAC. Now go out there and make me proud--dismissed!"

Okay, you now know from the previous Crewdog that those words were NOT going to be coming out of The Flame's mouth; I knew it, too. There he sat, flipping furiously back and forth between my different quizzes and check rides (trainers counted for this certification, also), and there was nothing I could do to change his disposition. This was my first encounter with him face to face. For some reason, my EWO certification had been given to the wing Director of Operations (DO); the Wing Commander must have been out of town or detained for some other reason. As he flipped back and forth in my training folder, I could tell he was becoming more and more frustrated. Then it hit me - he couldn't find anything to yell at me about, and it was driving him nuts. I couldn't believe that I was witnessing that, and I half thought of chuckling about it (to myself on the inside); however, under the serious situation, I didn't dare even chuckle to myself, even on the inside. Suddenly, without warning, he closed the folder, slapped it down on his desk, looked me straight in the eye with a look that would have melted a block of ice and said, "You'd better not f*** up!" He then waved me out of the room with a sweep of his hand, I saluted, took an about face, and departed after carefully closing the door behind me. On the bench outside sat the next poor soul awaiting his turn. I never slowed down or looked back until I got back to the squadron. I was leaving no chance for The Flame to call me back in for something he might have forgotten to tell me.

I had just experienced the operational Crewdog flying world of SAC in 1975, and it wasn't pretty for any crew position or for any crew. It probably wasn't even fun for the "staff weenies," either. Everyone was "under the gun" every day, especially for their bombing and electronic countermeasures (ECM) scores.

Even under this pressure, there were always humorous stories and events that showed the Crewdog's sense of humor in the face of these pressures on the flight line. Virg Love recounts two episodes that brought big grins to my face. "One week when we were on Alert, we were bored and tired of the guards at the entry control point sending us back around the Alert vehicle to recheck the wheel wells for bombs. So we put a B61 mock-up in the back of the truck. We drove out of the facility around the parking lot and when we got back to the entrapment area, we told them there was a bomb in the bed of the truck (a Nuke no less). They were not amused. That incident got us a face-to-face in our Blues with "the man" himself. It did not, however, get us kicked off Alert. Those were the days...."

Virg's second story goes like this: "One day John Caban was in the Command Post during an alert moving (elephant walk) exercise. The Flame was on the ramp to observe and things were not going smoothly. The Flame got on the brick and began issuing directives to John. The transmission reception in the Command Post was not the best; therefore, Caban replied "Alpha (the call sign for a Wing Commander) You are Broken" to which the old man fired back "I am NOT BROKEN. I am the Wing Commander!" and proceeded to throw the brick through the open driver's window of his staff car with such force that it bounced off the seat and up through the open passenger's window, shattering on the ramp when it landed. You gotalovit..."

The pressure intensified more and more, and soon we found ourselves preparing for an Operational Readiness Inspection (ORI). That would be the first one for most of the crews because they had been overseas for so long and out of the ORI rhythm of SAC prior to that.

Some guys, though, had been through it before Vietnam. By then, however, the mission profiles had changed due to the lethality of the Soviet Air Defense Systems, and we were forced to fly low to avoid being shot down on our way to the targets. As I recall, this transition to peacetime ORIs was so difficult that virtually every wing in SAC flunked that particular ORI route the first time they flew it. If I remember correctly, a bomb site near Little Rock, Arkansas was our ORI target area. On a humorous note, I recall that the Chief of Bomb Nav had a friend who owned a Cessna 172 and flew the route and took pictures so we'd know a little bit about what the target area looked like.

Senior leadership put a lot of pressure on all the crews to not make any mistakes. Bad bombs and bad ECM scores were like magnets to get you highlighted in the Quarterly Review Panel or Weekly Standup with the Wing Commander. Those panels consisted of all the Wing Op's staff and all of the squadron commanders. Yes, all - even the tanker squadrons weren't immune to it.

Glenn Burchard relayed his experience to me. Early upon his arrival at Carswell, he had been so good in Navigator Bombardier Training (NBT) at Mather AFB that he had been designated as a "baby radar Nav (RN)" from the very start. On one of his early flights, the crew threw a bad bomb. That was considered a team effort and no one person got the blame by himself. From that moment on, Glenn's crew was constantly harassed by the wing staff, the squadron commander, and those infamous "white-throated nit pickers" from Stan Eval (DOV). To quote Glenn, "I do remember after transferring to Carswell that as a still relatively inexperienced RN (certainly as it related to stateside low-level missions), I was put on an R crew with a new AC, Bill Hemmens; a new Nav (Sammy Saliba); and a new Gunner. We felt like we were being hounded constantly. Bo Fulford (copilot) and Lynn Wakefield (EW) were our only two experienced guys--that should tell you anything! We were no-notice check ride magnets!"

The day-to-day pressure was so unrelenting that eventually Glenn decided to resign his regular commission. As Glenn recalled, "The following day, the squadron commander, Lt Col Ron Beezely, called me in his office for a good conversation about his decision. I really didn't have any problems with him (Lt Col Beezley). When I gave him my logic for resigning and brought up the way the crew had been treated, he essentially told me that after I threw a bad bomb on one of my first missions, he was told to put the pressure on us and essentially use us as an example to put some fear in the rest of the crews. In fact, he told me in confidence that he was proud that we did as well as we did under all the scrutiny. I guess if it's stuck with me after 29 years it must have made an impression!" This indeed was an impressionable time for Crewdogs, and the key operative word was "fear."

In order to escape the horrors of those "Bad Bomb Panels," crews found ways to "play it safe." The pilots and navs worked with the schedulers to get them only certain bomb sites, and they would only go after certain targets and types of bomb runs at those sites in order to give them as much chance as possible for a "reliable" score. Also, the EWs would be sure that they requested certain types of air defense threats at these sites, and certain sites were almost taboo to fly against I remember the site at La Junta, Colorado being particularly lethal. Keep in mind that while we were "playing it safe" with "comfortable" sites and familiar targets. Our ORI, like our real EWO-certified assigned target area, would be against totally unfamiliar site targets.

What I do remember is that I was personally incensed that my EWO procedures in the AN/ALQ-T4 ECM Trainer were somewhat reversed by what I was told to do on a stateside training mission. I wasn't forbidden (except by specific stateside FCC regulations) to operate as if I was encountering real Soviet air defense threats; however, if I wanted to avoid any possibility of getting a bad ECM score, there was a way to cheat the system. I cannot describe the exact tactics due to classified restrictions; however, cheating the system meant that I would have to practice doing something in peacetime that in wartime would put the lives of my crew in potential danger. That just didn't sit well with me, and I argued about this with DONP (the Penetration Aids Chief) every chance I got. In fact, my silly pride demanded that I go against the grain of the status quo and use the same procedures peacetime that I would use wartime - even if it cost me and my crew a fateful meeting before the Wing Commander.

My AC at the time was John Magness, the CP was Ben Barnard, the RN was Dick Deroos, and the Nav was John Turner. Billy Bouquet was the gunner. It was my impression that the rest of my crew thought I was crazy to do this. They didn't try to talk me out of it, but they made it plain that if I "unsat'd" an ECM run, I was going to be on my own, which really was true. The pilots and Nav's couldn't be blamed for something the EW did. I sweated out a few ECM runs; however, I actually got pretty good at defeating the bomb plot sites. On several occasions, I would "tease" the threats through manipulation of my jammers. As I continued to defeat these threats, my confidence increased, too. Ironically, six months later, the ECM scoring rules changed back to the realistic way I had been training all along, and more and more EWs ended up with bad ECM runs due to the way they had trained in the air for so long. I think God protected me from my foolish pride.

For some bizarre reason, I didn't fear The Flame. I should have, but I didn't. Don't get me wrong, I respected his position and his power. In those days it was even more than just respect. It wasn't idolatry; however, you cannot imagine the power those guys had at that time. It dwarfs the power that Wing Commanders have today. The main point is that EWs, RNs, Navs, Pilots, and Copilots were really under the gun to "not make a mistake." That focus in training is what eventually cost the wing dearly on the upcoming ORI. I learned a huge lesson from the disaster that I carried with me the rest of my 30-year career. That lesson was: aside from safety of flight and nuclear command and control, in training, people have to be given room to make some mistakes, and then learn from them. Otherwise, learning does not take place, and they will fail when confronted with unknown situations. Give them enough room to experiment and to find better ways of doing things, to learn what works, and what doesn't without fear of being humiliated or jeopardizing their careers for the future.

The weekly "stand-ups" and the Quarterly Reviews prior to the ORI continued unmercifully. Virg Love remembers them well and said, "What a dog and pony show! Two dozen guys lined up in the hallway, with their respective documentation in three separate locations, on the slides, in the presenter's briefing book, and in the stack of Form 8 being handed to The Flame. Five Stan/Eval guys were on the firing line as much as the poor souls brought in for the event: one was in the projection room with the slides, one was in the hall with the victims, one was behind the podium with the briefing book praying everything was lined up (that would be me), one was on the inside of the door to prevent premature entries, and one was next to The Flame with the Form 8s. The potential for screw ups was infinite - wrong guy, wrong slide, wrong briefing sheet, wrong Form 8, or any combination of the above. Any error would send the old man into an epileptic fit storming out of the room. That's how I got the job. It's humorous now, but it was a frickin' nightmare every Thursday morning in the mid 70s."

Lynn Wakefield recalls a specific one of those vividly because he was flipping slides during this debrief (maybe that's how Virg got the job?). They were overhead transparencies for you PowerPoint techies of today. As Lynn recalls, "A radar Nav instructor came into the room to find a huge, long table. All around this table were the wing staff and the squadron commanders. At the end of the table was the Wing Commander, and to his right was an empty chair, where the RN sat. I read the details of the slide, including the words, 'Unsat for bombing proficiency; Unsat overall.' There was a huge wait, and at that point, nobody was breathing. You could almost hear the air leaking out of the door. Then The Flame said, 'Are you just stupid, or do you just not care?' There was at least an 8-10 second pause as I just watched him die. Then, The Flame launched into the RN and his squadron commander for a while. I remember five things resulting from that panel: the RN lost his instructor status, he was decertified for EWO, he was retrained for EWO, he had to do another re-qualification ride, and his crew lost their E status and were down-graded to an R crew." That was the result of just one of those weekly/quarterly panels.

Dale Fink, a nav, remembers his unlucky personal encounter with one of the panels. His RN had thrown a bad bomb during a mission that unfortunately was the only mission where Dale scored a perfect "Shack." The same routine that happened in Lynn Wakefield's story befell Dale and his RN, except there were two chairs up front by the Wing Commander. Dale's RN (name withheld for obvious reasons) got the same lose-lose question, "Are you just stupid, or do you just not give a sh**?" After another period of desperate silence and a tirade toward this RN, his crew, and his squadron commander, The Flame turned to Dale Fink, looked at his "Shack" score and said, "I've got nothing to say to YOU!" Boy, you can be sure that Dale was glad for that. Everyone was shaking in their flight boots by that time.

Well, as you can guess, crew morale was in the toilet. I remember at the time hearing guys brag that they were going to throw all of their bombs in the "boonies" during the ORI just to get back at the commander. I think most of that was just bravado and frustration. These were combat veterans, and they were proud of what they did over there. They were good at it, too.

It won't come as a huge surprise to you, but our wing busted the ORI. We were all in the auditorium as the scores came in and were posted. It was evident early on that things were not going well. Finally, it was obvious that we were going to flunk it. Spirits were low, and the Wing Commander let us have it with all barrels, even in front of our family members there with us. We all felt awful, but we couldn't change anything.

To improve our performance and morale, we were ordered to be at work every Saturday for the next six weeks to prepare for a follow-up Buy None (a self-run ORI). We studied targets, procedures, took tests, etc. It seemed that we knew what we had to do; we just didn't know how to do it correctly when it counted. Dave Lay remembers vividly how bad it got. He told me recently, "The wing had failed the ORI, The Flame was Wing Commander, and it was immediately after the IG left Carswell. I was taking my exams before my checkride in Stan/Eval test spaces on the second floor of Wing HQ building. The Wing Commander's office was on the first floor at the west end of the building. I was on the second floor in the center of the building in a testing room that was a good distance from the commander's office. All the crews that had bad bombing activity were being interviewed by the Wing/CC. The Squadron leadership/Wing agencies (CC, DO, Training Flight, Bomb Nav, Pen Aids, etc.) and each crew individually went before the Wing/CC to explain bad activity while flying the ORI. Sitting in a testing room, on the second floor, I heard the Wing/CC say from his office on the first floor at the end of the hall, 'Get the next leper colony in here and who owns the bastards.' I am certain the conversation went down hill from there. I was so very happy I did not have to fly that ORI."

The resultant Buy None was not much better than the ORI, and we went through the same routine, with the Wing Commander chewing out the crews in the auditorium in front of the families and then storming out of the auditorium. Only that time, I saw Brig Gen Burpee walking down the hall just before the Wing Commander had stormed out of the auditorium. The general had overheard what was said to the crews in the auditorium. Again, in that situation, all the Crewdogs were dejected and silent. Nobody knew what to say. To our surprise, we could hear yelling down the hallway that continued for about one minute. We knew who was out there, and every one of us could have heard a pin drop. Everyone was terrified as we saw The Flame walking back into the auditorium. Everyone snapped to attention as he came in and took the stage. He made an obviously forced speech about being wrong, and then he walked quickly back out of the auditorium. We were all shocked about everything and a little embarrassed for both him and ourselves. Within 24 hours, there was a moving van at The Flame's house, and within 48 hours he was off the base and gone forever. Those were times that were tough not only on the Crewdogs, but on the senior leadership also. After talking with Crewdogs at other bases and other wings, I have come to appreciate the fact that practically all SAC wings were run in a similar fashion, with each Wing Commander almost trying to outdo the next one in terms of stern rule and rigid training regimen.

The good news was that many of us Crewdogs learned valuable lessons from our experiences there at Carswell and at other bases, too. For instance, we learned that training is more successful when done in a way that allows people to learn from their mistakes and to improve or find new and better ways to do their jobs. That has helped me immensely in my later career in a host of training situations. Another lesson has to do with caution when mixing ego with power. I don't think that we ever will be able to avoid this problem in the military as long as there are folks who believe that "looking good is a full-time job." I've been on several bases where it was known that the base newspaper was told to have this or that person's picture in the newspaper at least twice in each week's issue.

Another lesson was that, in most situations, leaders should praise in public and admonish in private. It's no longer as common for Air Force senior leadership to berate squadron commanders in front of their troops. Finally, I learned that I shouldn't base everything I know about a person on hearsay. I heard a lot of stories from a lot of Crewdogs, and they seemed to be embellished more and more each time the story was told. I hope that my perspective has become a little better balanced through the exercise of writing this article.

For example, when I asked for inputs for this article, Bill Reynolds relayed the following personal story to me: "After I was passed over for 0-5 the first time, I tried to get some things changed in my AF records for the next board (although nothing got changed). I requested support from him (The Flame). He not only gave it to me, but he said in his letter that it was the first time he had chosen to do that. I have no way of knowing if there is any truth to that, but I was very impressed at the time (1988 or 1989 timeframe) that he would say that. Things have changed a lot in the AF...I have not met anyone like that in a long time."

Tommy Towery has his own opinions of leadership styles. He recalls that in one of the great classic air movies, *"12 O'clock High"* Gregory Peck is cast as Brigadier General Frank Savage. He is sent into a bomber squadron to clean it up and get the men in order because the previous commander has become too close to the crewmembers and did not show a strong military bearing. In trying to get the unit back into shape Col Savage is such a harsh commander that he alienates himself from all the crew force. As the movie plot progresses his men come to understand and respect the rough tactics of the new commander who in the end also becomes so attached to the unit himself that he is afraid to send his men into combat. Tommy recalls that there was also a movie called *"Dumbo"* about an elephant that could fly! Movie characters do not always portray reality and some real life people never change like characters do in the movies. Most elephants do not learn to fly and not every commander's tactics eventually earns him the respect and the love of his men. Such was the case in this instance.

Indeed, things have changed quite a bit in our Air Force, and there are many Crewdogs who will remember The Flame, or another Wing Commanders just like him. In the end, we probably have to admit that we learned a lot about leadership from them. Honestly, it was both good and bad, and we are better men and ex-Crewdogs for having gone through that unique SAC experience.

I extend my personal thanks to the following contributing Crewdog authors who helped with this story: Lynn Wakefield (Col, Ret); Dale Fink (former Capt); Glenn Burchard (former Capt); Tommy Towery (Maj, Ret); Bill Reynolds (Maj, Ret); Virg Love (Lt Col, Ret); etc.

Chapter Two

Cold War [kohld] [wawr] – *noun* - A state of political tension and military rivalry between nations that stops short of full-scale war, especially that which existed between the United States and Soviet Union following World War II.

Chrome Dome Missions
24-Hour Airborne Alert
Roland E. Speckman

By the time I had checked out as a 764th BS combat ready B-52D Aircraft Commander (A/C) qualified to command Crew E-06, SAC had flown over 2,000 sorties convincing the Defense Department that we should be flying these 24-hour Airborne Alert Sorties, fully armed including Nuclear Weapons. E-06 was selected to execute the next ordered 'Chrome Dome' mission when the 461st Bomb Wing next rotated.

This system of ready response at any hour had been perfected, updated, and executed under full wartime conditions. When Crew E-06 got orders to fly a long Chrome Dome mission, all of the needed details, timing, and flight plan information were pretty much canned. All we had to do was fill in the current changes and concentrate on the various target study folders. For the most in maximum security E-06 could only do their study and planning while on ground alert in the bomb proof alert quarters called "The Mole Hole." Wing policy dictated that all combat ready A/C's had to fly as an augmented crew member on one sortie to familiarize himself with all the problems he might face his first time out.

I completed this required trip in early January of 1965. We had been standing surface alert in one of six B-52Ds parked in the alert area close to the "Mole Hole." I was quite familiar with the day-to-day routine practices of the six day tour.

As the ever increasing threat of USSR's Intercontinental Ballistic Missiles (ICBM's) hitting SAC's dispersed B-52 Strike Force grew, SAC began hardening all alert facilities to at least protect the Quick Response Ready Alert Force. SAC Headquarters selected the most remote spot at each B-52 base that was still close to the runway to build these alert facilities. They were built to house the Combat Ready crews and support personnel. While the building was being constructed, so was the Alert parking ramp with room for at least six bombers and two tanker aircraft.

Plans for the newly designed building had the first floor completely underground and sectioned into crew sleeping quarters. Exits on two sides permitted crews speedy egress from the sleeping rooms by running down halls to exit up a sheltered tube to ground level. Some bases had an alert vehicle for each crew positioned near the exits. Other facilities were close enough to the cocked alert aircraft to allow the crews to just sprint out and get aboard the planes. The rooms needed for day-to-day training, target study, testing, eating, and lounging, plus office space occupied all of second floor. There wasn't a window to be found anywhere in the two storied alert working and living area. One thing I do know is at 1600 hours every weekday, all available crewmembers could be found around the lounge TV watching "The Yogi Bear Show."

With the surface alert task force in place, we were ready to launch prior to the arrival of any ICBM targeted at our Amarillo base. With 51 other SAC bases on ready alert loaded the same way, the folks in headquarters wondered if all alert B-52s could get off within the required 15 minutes time. They didn't think they could, so, the SAC planners got their heads together and came up with a 24-hour Airborne Task Force plan that used the same nuclear weapons, flight plans, maps, and targets that were used on surface alert. Areas being patrolled were controlled by SAC Headquarters to avoid inflight mishaps. In fact, when a B-52 that was fully loaded for its nuclear war mission came off surface alert, it would be assigned a flight crew and be readied for a Chrome Dome launch.

Crew E-06 had completed flight planning while on alert in the Mole Hole. With dates and routes canned by SAC relayed through 15th AF, and 461st Bomb Wing, all we needed to do was to update the plan with the latest information. Crew E-06's roster of A/C Maj Speckman, C/P Capt Montecino, R/N Capt Glick, EWO 1st Lt O'Conner and tail gunner MSgt Smith was augmented with an extra Pilot and Navigator. These spares were usually staff officers in need of monthly flying time. What better way was there to log 24 hours than all on one flight? They were of great assistance relieving crew positions for short periods of station rest, mainly sleeping.

Launch Day, 29 August 1965 06:30 hours

With a two-hour station time, E-06 was briefed for the last time and released to "Go do it." The crew, plus the two spares, piled into the crew van and headed for preflight. We had to onload crew, equipment, food (TV dinners and flight lunches), plus sleeping bags, jugs of water, and two large Thermos bottles of coffee - enough to last us for 24 hours. We were ready to live in the cramped quarters of the flight deck. But, first things first.

We hardly had enough room to roll out one down filled sleeping bag. One would think the big old BUFF would have plenty of space to move around in. From all appearances and outside views, there should have been plenty of room. By the time we stacked in enough food for the whole crew for the next 24 hours, the pilots had to run an obstacle course to get to their seats.

It was early morning, still dark, but near dawn and the pre-takeoff checks were completed when the C/P checked in with the Command Post.

"Home Plate this is Polar Bear One; checks completed; ready to start engines on your hack."

Home Plate replied, "Polar Bear One you are cleared to start engines 08:15 hours. Standby to copy ATC clearance." The enroute clearance remained as planned and was copied by the Nav, C/P, and EWO and repeated by one of the pilots."

Given a time check at briefing, the crew awaited the final call from "Home Plate."

"Start engines and cleared to tower frequency. Home Plate out."

We had been briefed to simulate a wartime mission realistically by not using the radios more than we had to; however, we were instructed to not compromise safety, especially during inflight refueling. We were told to try for a visual contact with radar approach and then use radio silence hook-up procedures for max security reasons.

Starting all eight Pratt & Whitney J-57-43w turbo jets using rapid start procedures, they were soon whining and just waiting to deliver 13,000 lbs of thrust per engine. We were cleared to taxi and moved slowly. On every turn, each pilot watched the bogey wheels under each wing tip. With a full load they were tracking on the pavement, but with the slightest turn error they could slip off of the hard surface miring down in the winter wet soggy ground. Thank Heavens all went well as we pulled out onto the active runway, rolling up on the yellow center line and listening for the "GO" hack. After being cleared for T/O from the tower, we didn't stop to set the brakes but continued on a rolling takeoff.

Over the interphone came the question: "Crew Ready for Take-off?" Station by station each responded including the spares. As the A/C advanced that big fist full of throttles steadily forward to take off power, the C/P backed him up to be certain one or more wouldn't pop back. We didn't have to use cross-wind settings on the four big gear trucks as the wind was on our nose. Our four main gears were astride of the center line and no corrections were needed. Line speed checks were made with more indicated air speed at each point until we passed the "No Go" decision line. Right at the air speed that was computed for lift off, the big bird came loose of the runway. Takeoff power settings were sending black smoky trails streaming out behind each of the eight engines almost blacking out the runway for the following aircraft. We were off, in the green, and putting the gear up. They slowly twisted up into their cubby holes and with a "thunk" were locked up with doors closed.

With the aircraft cleaned up and climb-on-course power set, the tension of takeoff disappeared. The nav, and radar called for a turn over Buffalo Lake, just 15 miles from the end of Amarillo's runway. ATC cleared us to climb to our cruising altitude of 32,000 feet and checked in to call upon leaving their control area. It was a beautiful day for flying. It was one of those clear winter days with the sun climbing up past the horizon and not a cloud in the sky. We just hoped that the first refueling track wound be cloudless. Of course, we let our fourth pilot, "George," do most of the level flying on autopilot.

Enroute Crome Dome" T/O + Six Hours

To take advantage of the selected entry points on the H-Hour Control Line our route was set to avoid populated areas, primarily because of our nukes. Where possible our flight would be over water just to be safe in the event of an inflight mishap. The navigator, and radar, with help from our EWO, maneuvered our BUFF around the course as planned so routinely that about all we heard from them was heading and altitude changes. The Copilot kept up with all the needed position reports.

Polar Bear One tooled on through cold winter skies as we began getting set up for a daylight inflight refueling with "Deep Six, a KC-135 tanker. During the next five hours we continued our radar navigation, avoiding population buildups, monitoring our four nukes, managing fuel to accommodate a maximum on-load, position reporting, and having a third cup of Joe to wash down an inflight lunch. There were a few discussions intermittently via interphone on current events such as Winston Churchill, Britain's WWII Prime Minister and world-wide recognized Allied leader, who had passed away just four days earlier. Without exception our crew wondered what a send off the world would hold for this "Bull Dog" warrior.

First Inflight Refueling:

We wanted to play the game as briefed regarding radio silence for our rendezvous through completion of this max weight off-load. We made our call to control, reported, and gave our on-time intensions. Before checking off, control reported that the KC-135, "Deep Six" was orbiting on station at altitude and awaiting initial radar contact to commence with the rendezvous.

"Roger that last, monitoring needed frequencies. Will not break radio silence except in emergency. Out." answered the C/P. The A/C asked for the Tanker Rendezvous Check List.

"R/N here. Maintain your heading 040 degrees. Start your decent to 1,000 feet below refueling altitude when we have radar contact with Deep Six." We have him orbiting the hook-up point. Will notify of his last turn to track."

We had already coasted out of Maine over the Atlantic getting our feet wet while headed for "Cold Coffee". There was ample time for the C/P to run the check lists.

The Gunner's report concluded this check list. The A/C was sweating the boomer out to set his boom in the receptacle while listening for that "thump" and the sound of the toggles latching. Constant attention to the tanker's directional lights was needed to stay in the boom envelop. It was a smooth day with little or no turbulence at refueling altitude. The fuel transfer was progressing rapidly.

Our fuel manager copilot worked over his fuel panel as selected tanks filled from the main fuel line. He also backed up the throttles for the A/C.

It was a sweat-producing job of holding the aircraft in position for the time it took to off-load such a large fuel load. It seemed like it would never get off the tanker and into our tanks.

I could feel the sweat go down my back and pooling around my butt. I do believe that was to be the longest time hanging on the boom I ever had to accomplish. I hated to think that I would have to repeat the refueling up over Alaska, and at night yet. I just hoped I might do it as well and complete the 24-hour patrol as briefed. It must have been about 15-20 minutes before the boomer disconnected from us.

Polar Bear One maintained position until the "Deep Six" tanker made a climbing left turn, reversing course to return to his home base. He had chalked up another successful max off-load and disappeared above and behind his receiver.

Nav corrected our course and altitude. "A/C come left to 030 degrees and start climb to 31,000 feet, your cruising flight level."

"Roger Nav. Course 030. Climb flight level 31,000." A/C repeated while taking the requested action. Nav went back to navigating and continued alerting the crew on the land masses we were flying over or close by so we might see some of that part of the world before the cold winter darkness settled in and around us.

After making a few entries on his fuel log, the C/P ran through his check list.

The copilot queried each crewmember for their responses which were acknowledged on the interphone when each item was accomplished. We went through each step while climbing on course and had it all done by the time we reached flight level 31,000.

I was feeling those hunger pangs after all the physical effort I sweated out while doing that max fuel off-load, so I asked for the spare pilot to occupy my seat while I fixed and ate my TV dinner. We changed places on the crowded flight deck. It felt good to relax and stretch whatever I could of my 6' 2 ½" frame. The best I could do was to put my legs down the hatch to the lower deck with my boots on the same step of the ladder and then straighten out my body by reaching for the sky then relax sitting there and repeat the maneuver.

The Gunner was good enough to heat up my TV steak dinner and I ate heartily. It all tasted so good, but then I had really worked for this repast so enjoyed myself while eating the whole thing. There we were patrolling as requested, everyone doing their job gathering all the information possible for SAC while listening for the "Sky King" messages. One of them could send our load into Russia or China depending upon which place we were closer to. We were ready but did pause over pencils ready to copy and authenticate the actual "GO" message. The Cold War had been heating up for the past few years to the point where we had to daily put up timed patrols to remind the Commie leaders that we did have an airborne strike capability up there to carry out delivery of four nukes. Beware!

Chrome Dome Mission Continues > T/O + 13 Hours

The priority task for all ferret flights was finding and plotting radar stations. The chain of interwoven sites kept forever increasing in their length and breath covering most of USSR. Soviet radar coverage became so powerful and cunning in tactics that U.S. Aircraft like our B-52 flights could record enough vital signs while skirting them over neutral territory. That proved to be us bombers. Those Soviet radar operators could set their watches on our daily arrival within their range. In fact, they wouldn't come up and track us until a few minutes before our turn away paralleling their coastline.

Our main interest on that patrol was activity in the Soviet West Front while we proceeded up near Thule AFB, Greenland, toward targets in Murmansk and some even down near Moscow. Swinging westbound over the Arctic wastes we would switch our attention to targets from Siberia down the Kamchatka to the Vladivostok corridor. We did cover a lot of ground to let the Soviets know we had the immediate capability of striking their vitals at any moment should it become necessary.

Sights along Patrol Route

E-06's regular Nav, had taken us this far, and was hungry and wanted to nap some before working on his celestial grid navigation leg. The leg had to be a minimum of two hours long to receive credit but the way the upper winds had been increasing we figured the leg would take longer than that. The Nav told us his intentions as he gave up his downward ejection seat to the augmented Navigator. Since it was more or less a do-nothing route up to the turning point at 60 degrees North (start of CGNL) he would point out things to look for while keeping us over the water all the way.

The adjustable lights of the instrument panel illuminated the eight-day clock as it ticked slowly around hour by hour. Except for alerting the crew to things outside to look for, things became pretty quiet on the flight deck. The witching hour was upon us.

We all had been briefed by the weatherman to be on the look out for a good display of the Aurora that night as we flew across the top of the world. He couldn't be specific but since we would be traversing the right area at the right hour (before midnight) we would have the best chance to get a good look at this phenomenon. He also told us a little about what caused the Aurora Borealis. The light displays are most active during magnetic storms and seem to be most frequent around the time of the greatest sun spot activity. The glows of frost-white beams seldom occur less that 50 miles above the earth but can shoot up as far as 500 miles, appearing to disappear into infinity. The most beautiful swirling glows were interspaced with ever increasing beams of indescribable colors - but predominately green. As they intensify in brilliance they bring in pale or deep reds. Ever climbing and intensifying curved bands of light blossomed above the glow with beams shooting up, up, and away. The flickering beams looked weird and frightening to one who did not know from whence they came. They change their brilliance second by second. When the Aurora reaches its full intensity it covers the whole sky with shifting curtains of light much like draperies wavering in the wind.

When the weatherman was finished with his briefing, I asked the crew if any of them had the experience of flying through an Aurora, or if they had seen a wondrous display of Northern Lights. Only the spare pilot and Nav had been through one. Right then and there we set up a plan of switching seats so everyone could have a look when we passed by. Then I quickly ran through the story of my first encounter with the mighty Aurora Borealis.

Nine years earlier when I was an A/C on a B-47 flying with the 9th BW out of Mountain Home AFB, Idaho, SAC Headquarters decided to launch two different simulated war time missions simultaneously. They were nicknames "Power House" and "Road Block". The exercises would last for two weeks and involve over 1,000 B-47 bombers and KC-97 tankers. It was in December of 1956. Our B-47 got off on time and proceeded with the directed requirements up to and including a night inflight refueling with a KC-97 TDY from Eielson AFB, in Fairbanks, Alaska.

We made rendezvous low and slow with that lumbering old 97 with a full load of JP-4 to off-load, heading right into an ever building Northern Lights display. We were hanging on to that boom for dear life. That KC was a black silhouette against the ever-increasing intensity of this swirling mass of colored misty glow and shafts of light. Since the KC-97 was not a jet it had to make a controlled decent on a verge of a stall to transfer his load to our B-47. That's where I almost lost it trying to stay in contact no matter what I thought the tanker was doing. I stuck to him like glue not wanting a disconnect until we got all of our fuel. I needed every last drop to finish the planned mission. By disconnect time, we were past the Aurora.

As the flight hours rolled by, we kept one crewmember napping in the down-filled winter sleeping bag rolled out over the flight deck. The pilot and Nav both had time in it then returned to duty posts. The bag was in so much use that it never was empty long enough to cool down. Flying through the dark and black Arctic skies found me studying the multi-million star-studded-sky seeing things imagined or real, like shooting stars or other planes. I did not see another soul or aircraft except the tankers. I would swear we had been following a plane with its position and navigation lights on. They looked like they were not only flashing on and off, but also changing color.

"Crew R/N, I'll be starting my celestial grid nav leg at our next turning point. Gunner, please help me record my sextant celestial shots, while the rest of you scan the horizon for the Aurora." The crew responded by their station on the interphone.

Better than half way up this navigation leg, those on the flight deck could faintly see the starting glow. Others were called up to gaze to our right and slightly behind our BUFF. It looked like our navigation team had put us well ahead of the Aurora. We would miss flying through the beautiful show of Northern Lights. It was far enough away not to fly through it and yet close enough to give everyone a good look. It was certainly something to behold. It was another war story to tell the grandkids someday if we lived that long.

With all requirements completed to get full credit for the celestial grid leg, the Nav asked for a course change to coast-in on Point Barrow, Alaska. He shaved off a few miles by doing so and gave us an ETA to the second in-flight night refueling area of "Cold Coffee". We would arrive and rendezvous on time. Approaching the Alaskan coast we could just make out a few dim lights flickering in the villages down below.

Our fuel manager (C/P) had been keeping a good center of gravity and was getting ready to run the refueling check list when the A/C requested it. We had been splitting nap time in our seats. I can say without a doubt that the most boring part of this long patrol mission was that of monitoring instruments, while letting George fly most of the time. Of course any course changes needed were executed immediately by a twist of the autopilot turn knob - real hard work. Also, we split up the pilot's flying time three ways. Alaska was dark and snow covered. We continued to pick our way down towards Eielson AFB seeing larger concentrations of land lights marking the increased population living down there.

Our Nav with the help from the EW reported our tanker "Frigid 25" was orbiting on rendezvous point and had acknowledged our presence and would affect a join up on time.

"A/C to Nav. Roger your last. Will be keeping a look out to make a visual on him. Just let me know range and bearing on his roll out on refueling track."

It was a perfect night for this heavy off-load of fuel. We should be able to stay with "Frigid 25" and get the transfer with one hook-up.

"Crew from C/P. I'll keep my fingers crossed hoping you can. Ready to run the refueling check list for the last time this trip." Everyone was awake for this critical refueling success.

Night In Flight Refueling In Cold Coffee Area

Although feeling the effects of the 15+ hours logged, E-06 was alert and on their toes for a text-book manual rendezvous and contact with the super KC-135 at altitude and airspeed. The crew was surprised but well satisfied that we got our scheduled off load just about the time we overflew the lights of Eielson AFB. All had done their part just perfectly, adding to the successful fuel transfer that guaranteed the remaining nine hours of flying time needed to get back home. Upon completion of the post-air refueling check list, the EW and Nav got things going again to provide a new course heading and altitude. As A/C, I turned right to a westerly heading that put us heading close to the H-hour control line for Siberia. It wasn't too long before the EW informed us that we were being tracked by coastal radars. He started recording the different signals as the R/N made sure we didn't get to close to the line or to the coast for that matter. It seemed like the Soviet radar people were perfecting their tracking techniques on the Chrome Dome flyovers, but to my knowledge they never sent up their fighters to intercept any of us just as long as we didn't violate their air-space. As we flew along the planned patrol route the Russian radars would hand us off to the next radar area even before we really left their area of responsibility.

That went on along our southwest route, as well as the turn back to Alaska. It continued all the time, and for the whole flight for that matter. We all monitored the "Sky King, Do Not Answer" calls hoping against hope that we wouldn't get that "Go" message. Of course the farther south along the coast we flew the more intense their defenses became. The R/N and EW team were very busy listening and plotting so we could keep abreast of the H-hour entry points that determined which set of targets was ours to hit from a point of crossing. We were all looking for that last point before turning back to the USA. Even doing so, we were still held responsible for other entry points along the H-hour Control Line that we could get to and to the targets assigned beyond, if we had enough fuel remaining. So we were all very interested in the C/P's report when we reached that point where we couldn't reach anything but Home Plate with fuel left. We really were not out of business even then for I am certain that if tankers were available SAC would have arranged a fill-up and given us a new entry point to go to, depending on how badly our tanker forces fared on any initial ICBM strikes.

The R/N started painting some land masses that were mostly islands of the Aleutian chain off the right of our course as we flew into Anchorage. That was our planned turning point for our return route to the Southern 48 States. Everyone who needed to use the sleeping bag had done so for the second time. I was holding off so I could be the last one in it, maybe an hour before landing. However, I did get to eat whenever I got the hungries. The food supply was dwindling to almost nothing; and the coffee was all but gone. Whoever wanted a hot drink from one of the ration boxes would have to heat up a cup of H_2O in our hot cup and stir in the concentrate coffee or cocoa before he could enjoy that comforting warm feeling in his gut.

Crew All Tired But Pressing On

With the most hazardous part of our patrol completed and with a heading and course set for Amarillo AFB, Texas, we all began to come alive and to wonder just what time of the day we would touch down. Ah! A good hot shower, and sleep for at least 24 hours. The EW reported that his signal hunt seemed to be for naught and was ready for a nap, so as soon as we got in towards Alaska where the R/N could control our course, he did just that - went beddie-bye. He had to work hard over his equipment plus the effort of logging info and deserved a break. I don't know how those guys on the lower deck could stand their dark dungeon for hour after hour. They told me that as long as they kept busy those things didn't bother them.

I often wondered just what good these armed 'Chrome Dome' sorties were really doing besides training the combat ready crew to fly 24 to 27 hour missions. One of our flights once ran into some strong winds that put them in the "Cold Coffee" refueling area with minimum fuel. His tanker gave him every last drop to where the tanker had to declare an emergency and had to ask Eielson for a straight-in approach and landing. That's cutting it too close, don't you think? For one thing - ever since SAC started airborne alert it kept Primier Khrushchev from making any more threats. Oh well, all of us SAC professionals collect our pay every month for doing as directed and living up to SAC's motto - "Peace Is Our Profession."

Anchorage was cloud-covered, but since we were cruising on top of the clouds and the closer we came to its city lights the more the glow spread and penetrated the lower layer, growing larger but keeping the shape of the city. Since that was our next turning point, we did not announce to the crew what we saw. The pilots wanted to check the Nav team's turning point call. They made the call just as we were over the center of the glow of city lights. In passing we thought we could see off to the north the highest mountain on the North American continent, Mt. McKinley, towering up to 20,320 feet. The R/N said he doubted we could see that snow covered giant since it was 150 miles up the pike. He thought we saw the top of a cloud off in that direction. The R/N gave us our new heading SSE bound and we rolled out of the turn setting course for home plate direct.

Much of the rest of this patrol was routine - hours of boredom punctuated by minutes of sheer terror. We coasted in over Southern Canada near Vancouver and soon entered the good old US of A. Of course the Seattle area was cloud-covered, but in passing we knew that Fairchild AFB, Washington, was near at hand in case of any unforeseen trouble we might get into. That old bucket of bolts had held together so far, so was a few more hours flying time to much to ask? Oh! As we passed Mt. Home AFB, Idaho, I switched to Command Post (9th BW) frequency asked for and received a phone patch to my B-47 R/N, Major Tom Casey. A sleepy voice soon was on the horn asking what time of day it was. He was surprised I called. He wasn't sure where he would be going since the 9th was retiring all of its B-47s and replacing them with B-52s.

30 Aug 65, T/O+19:45 Hours
Lost Two Engines Just Short Of Amarillo

After talking with Tom Casey, I thanked the Command Post and switched frequencies. I got my spare pilot up in my seat so that I might grab a few winks before committing to the instrument flight, approach and landing at Amarillo. Weather reports indicated frontal passage in that part of Texas was expected about the time of our ETA. The cold front was moving quiet fast so we were hoping it would be past Amarillo AFB before we had to start our decent check list.

The sleeping bag was still stretched out on the flight deck and looked inviting and felt the same as I snuggled down into it. I must have been really pooped and I was asleep in nothing flat. I was out cold; didn't even hear a thing for an hour or so. I was wearing a headset listening in on the interphone. About the time I was supposed to get up and resume my A/C's chair the two pilots were busy shutting down the number five and six engines. Engine instruments indicated that a blade had been tossed - going through the twin engine hanging along side of it. They were so unbalanced they vibrated the BUFF horribly. Capt Montecino, E-06's C/P, happened to be flying the big bird at the time of the incident. He completed all the shut-down procedures on both engines before I could get back in my seat and strap in. With all things under control contact was made with our Command Post relaying our problems to the duty officer.

We held our course into Amarillo AFB, but heard they were below minimums for our approach and landing. We were advised by our Command Post that we were to proceed to Carswell AFB, the closest alternate that could secure our four nukes. Sure enough, the weather front had passed that part of Texas and except for a few scattered, scuddy, low clouds, the ceiling and visibility were both well above minimums. Approach control took us in tow and began vectoring us around to the beginning of our instrument approach. We decided with our gross weight we would make a high visual approach keeping our speed up till over the approach end of the runway then make a fly on touch down before pulling the power off. It came off without any difficulty. It was something to have all of the base's crash and rescue vehicles tearing down the runway slightly behind us. We could still taxi the B-52 to a security area by following the Follow-Me truck. He parked us and before shutting down the other six engines, the security force was putting up rope all around the nuclear bomb-ladened BUFF. That was done to control all outgoing and incoming people and enforcing the two-man policy within the area. We unloaded our bags and baggage not knowing just how long we, as a crew, would be staying at Carswell.

We checked in with the Command Post, were debriefed, and had to hand over our target folders and other sensitive information before they would let us chow down and get a good night's rest. One of the crewmembers needed something from our BUFF's flight deck, so I went back out to the plane with him. The guard out front gave us a little trouble for he had been told not to admit anyone up into the plane.

We had the security guard use his radio to call the Sergeant of the Guard to come out to his post and confirm that we were, in fact, the crew that had landed and parked the crippled B-52 loaded with four nuclear weapons. In a short time, a blue security jeep drove up and, in turn, was challenged by our guard holding us at bay. Boy, I've never seen security so tight around a B-52 parked on the ramp, even though it was in a maximum secure area and posted as such. We were afraid they might spread-eagle and search us.

After the sergeant gave the proper password he was permitted to get out of his vehicle and was recognized by our guard. He was carrying a clipboard with our crewmembers listed. Still, he checked each of our ID badges against his list before permitting us to board the BUFF to clean out our own belongings. The guard watched us unload so as to confirm that none of us would get near the bomb bay and its highly destructive nukes. He and we were all relieved to get away from that area. As we left the ramp we noticed that wing maintenance personnel had unbuttoned the cowling about five and six, and were getting ready to drop the two engines to be replaced. You can say that about the SAC team they all seem to be "Gung Ho" about doing their part to get that bird back in the air ASAP.

Accomplishing all of these regular "Chrome Dome" airborne alert missions was not without incidences good and bad. At the same time, flying them did delay and eventually eliminate the threats and ambitions of world domination by the USSR.

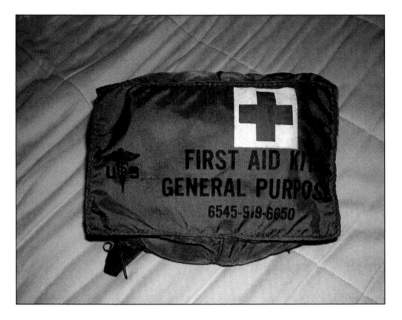

Chrome Dome Dis-asster!
Bob Stewart

My first B-52 airborne alert flight was most memorable. It literally was a near dis-asster to say the least. Our crew, commanded by Major Willie Clark, was the newest formed combat ready crew at Glasgow AFB, Montana. This was in 1963 when SAC kept a certain number of B-52 nuclear bombers on airborne alert in addition to those on ground alert. The missions were usually about 22-24 hours in duration and were flown around the polar cap. They were called Chrome Dome flights and were assigned at various intervals to all B-52 wings.

On all Chrome Domes crewmembers were required to fly as if they were on their way to their emergency war order targets. Real nuclear bombs were loaded in the bomb bays. Crews had to reduce cabin temperatures drastically so they would not fry in their heavy insulated winter gear required to reduce the shock if bailout over frozen terrain became necessary.

As the EW it was my duty to monitor the HF radio for the usual Sky King messages and possible receipt of the go-code to proceed to our prescribed EWO targets. Crewmembers always hated when they had to help monitor the HF as the periodic messages had to be decoded and authenticated, plus there was usually a lot of static and noise associated with it. The pilots and navigators were usually involved with air refueling and mission instructions from the two UHF radios while the EW was monitoring several radar warning receivers plus the HF radio. Radios and go-codes became drudgery after the first 20 hours or so on Chrome Domes, unlike the shorter 11 hour training flights.

It was sometime in the late stages of our first Chrome Dome that I informed the crew I was going to the lower deck for coffee and ice cream. No one else asked for coffee, so I was free to help myself to refreshments. I noted the radar navigator had an electric skillet plugged into a socket with a lid covering boiling water. He was doing that to thaw a frozen packaged dinner. The skillet was setting on top of the freezer chest which was filled with dry ice to keep the ice cream sandwiches from melting. To get access to the chest to get my ice cream, I picked up the electric skilled and sat it behind me on some gear.

Did I mention we were wearing heavy arctic winter gear? This made movements rather cumbersome at times. Well, while I was digging around in the freezer chest, my rear end came in contact with and tilted the electric skillet handle. That caused the boiling hot water in the skillet to spill and to saturate my very thick winter flight suit and thermal underwear. The multiple layers of clothing prevented me from feeling the boiling water at first. But after a considerable amount had been soaked up by my clothing, I felt the searing pain! My initial reaction was like the time when I was stung by a nest of yellow jackets. At first I had no idea what was causing the pain, which became more intense with the passing of each second. Soon the pain covered my entire rear end.

When I finally realized that it was scalding hot water that was causing my pain, I moved away and attempted to remove my clothing from my butt. Did I mention I was wearing heavy winter gear? By the time I reduced the contact with the hot wet thermal lining, the skin on my butt and legs was thoroughly saturated and ready to peel off. Not only was I severely wounded, but I was embarrassed almost beyond belief. How could anyone have been so clumsy and stupid to scald his butt on a B-52 flying at 41,000 feet on a Chrome Dome flight? That was what I was thinking when I realized I had to notify the crew of my predicament.

Meanwhile, both the navigator and radar navigator who were sitting a few feet from me had no idea of what had happened. As I tried to explain to them off interphone what had happened, the copilot announced that he thought he had heard the go-code on HF radio! Wow! That grabbed the attention of the radar navigator and pilot, so I got no help from the RN. The navigator was too busy navigating to assist me. Finally after determining there was no valid go-code, I managed to climb the ladder back to the celestial navigation area on the top deck and lay on my stomach. The pilots could not believe my predicament. Only after the RN said someone should look after me did the pilot order the copilot to come back and see how badly I was scalded. His reaction was that it was very bad. But he pled ignorance as to what he should do. He was most certain that he was not about to touch my posterior in any way, shape or form!

After polling everyone except the tail gunner, Willie Clark decided he would be unable to coerce anyone on the crew into administering first aid to my posterior. Therefore, he had to take two important actions. First, he would have to break open the first aid kit and obtain the salve, gauze and tape needed to medicate my wound. Second, he had to decide if higher headquarters should be notified of the accident and what, if any, decision should be made about continuing the mission. It was obvious that I could no longer sit in my ejection seat especially with a heavy parachute strapped to my back. Second, I could not monitor and operate electronic warfare equipment to protect the bomber and our crew should a valid go-code be received. So Willie had to ponder the ramifications of making a wrong decision about my condition, the ability to complete an EWO mission and the successful completion of the Chrome Dome flight.

The final decision was mostly no decision at all. Willie personally acted as doctor and nurse since no one else would. After threatening his crew about spilling the word that he had nursed my posterior back to health, he instructed all to proceed with the mission as planned. As it was, it turned out all right, and I managed to lay on my stomach the remainder of the mission. SAC never knew and therefore never banned the use of electric skillets on B-52 aircraft. Crews were spared from special training concerning medical treatment of burn wounds on arses.

With admonishments to the crew, Willie assumed the flight surgeon was the only other person who knew of the incident. That is, until one night at a party at the officer's club he overheard laughing and comments about an unusual relationship Willie Clark had with his EW! Sigh! Willie never found out which of his crewmembers spilled the beans. Needless to say, we all lived to see another day and to fly another Chrome Dome, but never to suffer another dis-asster.

Turner AFB, GA Crew E-24: SSgt Peter David "Scotty" Burns (G), 1/Lt Derek H. Detjen (EW), 1/Lt David McComb (Nav), Capt. George Benoit (RN), 1/Lt Dale Kuhns (CP), Major Theodore "Ted" Heiland (A/C)

A Memorable Chrome Dome
Derek H. Detjen

My first-ever "orientation flight" in the B-52D at Turner AFB, Georgia, was as a non-combat ready spare on an airborne alert mission during the height of the Cold War. On 1 Nov 62, Major Fred Robinson and crew slipped the surly bonds for a cool 23hrs and 15min of flying time, doing three trips around the Mediterranean in the process. On my very next Chrome Dome, on 12 April 63, the loss of one engine and the imminent loss of a second one dictated an air abort into Moron AB, Spain. Naturally, no one on the crew had any civilian clothes, so it was off the BX the next morning to purchase the standard "ugly-American" outfits of slacks/shorts, sport shirts, tennis shoes, etc. We spent several lovely days in downtown Sevilla, marveling at its lovely parks and gardens and the immaculate cleanliness of the whole city. Little did I envision that more excitement lay ahead in our Cold War standoff with the Russkies!

A mere month later, on 5 June 63, I made a mistake that I vowed never to make again, taking off in the IP seat on another Chrome Dome. Departing Turner AFB with my crew E-24, commanded by

Major Ted Heiland, we lost an outboard engine after we were committed to takeoff, staggering into the air at maximum gross weight, barely skimming over the trees on the Flint River as we fought for airspeed and altitude. The Command Post immediately directed us to abort the mission and after burning off enough fuel to attain a reasonable landing weight, we landed, logging only five hrs and 25min. The ground crew found a large pine tree branch in the number one and two engine pod, moot testimony to just how close we had tempted fate! I can still remember the view out the cockpit windows, making a note to avail myself of my ejection seat in the future. Major Heiland was to soon play a part in our most exciting Chrome Dome ever.

Theodore R. Heiland was a pilot from the old school. I remember seeing a picture of Ted in his leather flying helmet, standing in front of his Stearman bi-plane while attending Tex Rankin's flying school, I believe, in California. Ted knew the avionics of the D model by heart, fuel transfer sequences, electrical systems, etc., without having to resort to the Emergency Procedures Section of the Dash 1 in most cases. He was, and is to this day, the finest and most professional aviator I ever had the good fortune to fly with. His superior airmanship would stand him in good stead sooner than we realized.

After a lull of five uneventful Chrome Dome missions, our most harrowing one began on 26 Jul 65, departing Turner AFB in aircraft 5-00596 at 0811 Local Time. By then the Wing had seen too many air aborts and landings in Spain, so our erstwhile Squadron CO had dictated that "no civilian clothes would be taken on future airborne alert missions, much to our consternation. The flight went uneventfully until we were less than a minute from completing our final refueling in light turbulence. A sudden updraft caught us unaware, and before the boom operator could disconnect the boom, it reached the inner limit and smashed through the refueling receptacle, causing an immediate explosive decompression as we began our emergency breakaway. There was a large amount of fuel inside the crew compartment, due to the scavenge pump working in reverse. As the crew went on 100% oxygen, I was directed to take a walk-around bottle and inspect the damage. Peeling some of the insulation away from the roof of the fuselage, I discovered the damage while being soaked in JP-4. Major Heiland declared an immediate "Mayday," and Barcelona Approach Control cleared us for an immediate straight-in approach and landing at Moron AB, Spain. We touched no electrical switches, fearing a possible explosion in the crew compartment. Within a minute or so, we had a phone patch to the chief Boeing test pilot in Seattle, Washington on one UHF radio, and SAC HQ on the HF radio. The landing would be the heaviest ever attempted in a D-model, looking much like the space shuttle!

Due to our almost maximum gross weight, the flaps-up approach speed was 235 knots (about 270mph) with about a 25 degree nose-up attitude. Although Major Heiland offered all of us the opportunity to bailout prior to landing, we elected to ride the bird in with him. With a lot of help from Boeing, Ted made a literal "grease job" of a landing, as we used up the entire runway, shredded the drag chute and burned up most of the brakes before ending up in the runway overrun. We exited the aircraft in a rather quick hurry as the fire trucks took over. As memory serves, the landing weight was about 441,000 lbs! For his superb handling of the emergency, Major Heiland was awarded a richly deserved DFC! Many other close calls would follow, especially during the Vietnam years that would reinforce my original perceptions that the B-52D was one of the most forgiving aircraft ever built.

As a pleasant footnote to this mission, we managed to time our arrival at Moron to the La Feria Festival in Sevilla (much like our Mardi Gras) and after purchasing another wardrobe of "ugly American" outfits, were able to observe the running of the bulls at first-hand!

Carswell Quips
Ken Schmidt

Here are a few instances I remember from the Cold War era of my career that I fondly call my Carswell days.

My first crew was R-62 which later became E-62. The crew had a great pilot team with John Chapman as the AC and Joe Schwab as the CP. We were on alert one day and we were outside looking at the transient area near the alert pad. Being an avid airplane fan and plastic modeler, I mentioned "see the interesting red and white markings on that F-4 over there". They both said "What F-4". Both John and Joe wore glasses and I guess that I wasn't concerned about their being able to see while flying. One was farsighted and the other nearsighted so one could make the approach and the other one the landing.

On another alert tour, we were returning to the alert pad after a night Klaxon and aircraft response exercise. While going up the ramp, I found a live M-16 round on the ground. I picked it up and gave it to the chief cop in the Security Police office. I mentioned that "I think one your guys is missing something.' I never found out what happened to the cop.

On another alert tour there was work going on at the far end (south end) of the alert parking area where a large "bionic gate" was being constructed. On the north end of the alert area was the infamous "red line" which no one was allowed to cross (except in response to a Klaxon). Two cement trucks were parked just shy of the red line and one proceeded to cross the line without permission. He did not get far before he was pulled over, removed from his truck and forced to assume the position. His buddy (who fortunately had a little more common sense) did not follow him but was heard on his radio calling his dispatcher: "They got Leroy. He's all jacked up. Get him some help." Never a dull day at the alert pad.

There was also the famous "entrapment area" going into the alert facility. It was a large fenced area with two sliding fences in which one would open up to allow the alert vehicle to be driven inside and then the fence would close in effect "trapping" the vehicle in order to inspect it before allowing the vehicle to proceed through the second, secure moving fence. One crewmember would drive the vehicle through the entrance inspect the vehicle with a security policeman, exchange his line badge, then drive the vehicle through the other gate. Other crewmembers would enter a separate gate, proceed to the entry control point to exchange their alert badges, and then go through the turnstile to get onto the alert facility grounds. This whole process was designed to deter terrorists, but all it actually did was hassle the alert troops. If essence, if there was a dedicated terrorist, would he really go through the entrapment area or would he just proceed across the red line and drive his vehicle directly into a fully loaded bomber?

Get Off My Airpatch!
Bill Jackson

In January of 1968, the world learned that the North Koreans had captured the Pueblo spy ship and were parading the crew in front of the public. Part of my daily routine as a "Wing Weenie" in the 91st Bomb Wing at Glasgow AFB, Montana, was to go to the command post and pick up the mission paperwork from the previous day's training missions. The command post had those missions posted on a large blackboard with the actual takeoff and landing times. One B-52 sortie had been scheduled on the night before, but there was a line drawn through its entry with the words "Cancelled by SAC" in the remarks section.

It soon became known that the wing had been alerted for deployment to an undisclosed destination. The alert force was downgrading, and the schedule for the departure of aircrafts and crews was being drawn up. The wing had a couple of BUFFs that were not properly equipped for Arc Light type operations. That weekend those aircraft were replaced by two others from different wings. I don't think that the Crewdogs were ever briefed about a sudden departure, but it was taking place in an orderly pace just like the plan specified.

I am not sure of the dates but I do know that on the last day of the BUFF departures, only two aircraft and crews remained on the base. The Wing Commander, Col George Pfeiffer, and the vice Wing Commander, Col Hutchinson, who was on orders for an overseas post, were also still present on the air base. Col Hutchinson had volunteered to accompany the wing on deployment to take some of the load off the commander's back.

I was on orders for transfer to Beale AFB, and I was at my desk wondering about what I was going to be doing for the next few days with no planes or crews around. About that time an airman came in and said I was needed in the hallway. There was a hall that ran from the east side of the so-called "Head Shed" to the west side with an exit door on both ends. My section was to the north of the hall on one end and the commander's office was on the south side at the other end.

I walked to the hallway in time to greet a Lt Col and his 15 or so team members, all in flight suits, white scarves, caps and polished boots. He announced that his team was there to conduct a no-notice Combat Evaluation Group (CEG) evaluation of the 91st Bomb Wing. He asked me to escort him to the Wing Commander's office, and he also wanted copies of the weekly flying schedule.

I asked him if he had noticed the fact that only two BUFFs were on that huge parking ramp. I told him there was no schedule available and for him to follow me. I led him to the CO's office where I announced his presence, and then I stepped back to watch the fireworks. Without taking a breath and in a loud voice the boss stated that he had been praying for years for this opportunity!

He stated, "If your transport is still on the base, you have 30 minutes to get off my airpatch."

As the Lt Col departed I heard him say to his staff, "Why didn't SAC tell me?"

About 10 AM I was in the DCO's vehicle watching the second BUFF lifting off with their destination being Kadena Air Base, Okinawa. All that was left on the ramp were two fire bottles and two sets of chocks. Somehow every piece of equipment, tools and personnel had been airlifted overseas by tankers and MATS aircraft, most of which departed during the night. I heard Col Pfeiffer remark that he didn't think it was possible to deploy an entire bomb wing successfully in such a short period of time. The logistic plan worked!

Glasgow AFB eventually was closed. The former base is now named St. Marie, and many of the housing units are for sale. My first quarters were used as the sales model by the first purchaser of the base quarters. Boeing has purchased the runway and flight line. The runway was used for the crosswind landing test flights of the Boeing 777 among other things. For all of us who were once stationed there we will never forget the role it played during the Cold War.

B-52 Operation Port Bow
at Kadena AB, Okinawa
James Bradley

The capture of the USS Pueblo by North Korea, in late January of 1968, resulted in the 91st Bomb Wing making a unit move to Kadena AB, Okinawa on 28 January 1968. The crew to which I was assigned was flown to Kadena AB via a MAC C-141. When we landed and were taken from the C-141 to a crew briefing area, we passed several nose docks where strange airplane tail sections were jutting out from the doors. These were A-12/SR-71s, although no one would agree that this was what we saw. They were A-12/SR-71s! There was no doubt. We received a situation briefing and target and threat briefings and were sent to standby for further instructions. After about two weeks, there was the need to "Get them busy." That is how we started bombing South Vietnam from Okinawa.

We began flying missions from Kadena AB to South Vietnam on 17 February 1968. These were typically missions of from seven hours 50 minutes to nine hours 10 minutes. Once while flying out of Kadena AB, the Navigator, Bill Redman, was talking with the Okinawan who was driving our crew van. Some protesters were demonstrating against the use of Okinawa as a base for bombing in Vietnam. They were at the fence at the west end of the runway. Bill asked him if he demonstrated like that and the driver answered that he did. Bill asked him why he did that and at the same time he drove this van for the U. S. Air Force. The driver's response was, "They (the protesters) pay me."

While at Kadena AB, the Pilot, Copilot and I, played golf on the Kadena AB Golf Course. This was one course where, if the ball went into the rough, you left it and used a new ball. There were poisonous snakes that lived in the rough, which were called "habu." This also happened to be the nickname/call sign for the SR-71s stationed there at Kadena AB.

Some crews were deployed to U-Tapao AB, Thailand, sometimes referred to by its TACAN call letters, "BUT." After about two weeks at U-Tapao AB, crews were either rotated back to Okinawa or to Guam. Missions flown from U-Tapao AB were short when compared to the 12 hour missions flow from Guam. A typical flight from BUT to BUT was in the range of three hours 10 minutes (3+10) to four hours and five minutes (4+05).

On 24 April 1968 the crew to which I was assigned, returned to Glasgow AFB, Montana, via Anderson AFB, Guam. When I landed at Anderson, I had logged 2,356 hours in the B-52D. When I arrived at Glasgow AFB, I immediately began preparation to go to Southeast Asia in the F-105F aircraft as a "WILD WEASEL" backseater (EWO). Suffice it to add that as a WILD WEASEL Bear I had the opportunity to spend 60 days TDY/Enroute to Southeast Asia at Kwang Ju, ROK Air Base. The reason was to augment the 12th Fighter Wing at Kadena due the Pueblo's capture. I lay claim to being involved in the USS Pueblo's aftermath from both a B-52D EWO's point of view and that of a backseater in the F-105F WILD WEASEL.

[Please note: This story compliments Col Bill Jackson's Secret Deployment story]

Alert Facility at Guam.

Flight Suit Load
Walt Marzec

This is a story about one very non-typical day in my life as a B-52D crewmember. I was the electronic warfare officer on my crew, and we were deployed to the beautiful island of Guam in the Philippine Sea during the fall of 1973.

It was a particularly odd kind of a day - the sky seemed different, somehow. Maybe in hindsight it was different, or maybe it had something to do with that "whiskey fog" that had blown in during the previous night. Who knows? As I recall, there was no flying scheduled for that day, which was in itself unusual. We were all in what was called a "stand-down" condition but I didn't know why. I had rechecked the schedule, and sure enough, nothing was flying that day.

My aircraft commander was ordered to attend an afternoon meeting and returned to tell us what he was allowed to tell; which, as it turned out, wasn't much. He said the bomber and tanker forces were all in crew rest, and as he passed out the six-pack of Bud, he said, "The next time the Klaxon goes off, it's the big one, Martha." The R/N, being a Mormon, never drank; however, on that day he knocked that Bud back like it was a can of Coke and he was in the middle of a very dry place. The A/C went on to say that we were being held in reserve, sort of like a second wave, should we be needed. That had an ominous feel to it. I remember thinking, "Reserved for what?"

Meanwhile the commander came in. He and his wife were living just off base. They had just gotten married back in the states and he couldn't bear to be apart from her for six months. Lucky him - the Guamanian chicks were not exactly what you would call attractive. Anyway, he got to the BOQ and asked what was going on because some of the airmen riding on the bus he had come in on had asked why they were uploading nukes.

I also remember that evening vividly. I was lying on the bunk reading, of all things, a book about Israel's Alert force - strip alert for the Mirages they were flying. Anyway, Guns was standing at the foot of the bed, and we were generally BS'ing and speculating as to what this all was all about, when the horn went off. The look on the gunner's face was one of stark terror as I recall. He was probably thinking the same of me. I remember it like it was yesterday, looking at the gunner and saying; "Well, let's go watch." Our crew emptied out of the rooms, and all the other crews poured out of their rooms as well. We finally all congregated on the rails of the second floor. Our BOQ was the closest one to the Alert Facility, so we had a clear view of the entire production that was unfolding. Guys were streaming out of the facility jumping in trucks and gunning it to the flight line. Everybody standing at the railings began yelling and cheering the guys on as they jumped into the trucks and took off. We watched as they got to the flight-line, started the engines, and then started to taxi (Tanks and BUFFs alike). I was convinced that "this was it."

Then, after a half hour of this drama playing out, the command post chief, a Major I think, came out and yelled to everyone standing at the railings that the Klaxon had malfunctioned, and to disregard it. At that point, it might have been a bit hard to disregard a "flight-suit load," though. Anyway, we all stood around for a while watching as everybody taxied back. Somebody later (I think it was the next day) said that somebody had even jumped off the second floor balcony during the previous evening's festivities and had broken his leg. I don't know what that was all about. After a while, everyone drifted back to their quarters, probably wishing there were some more Buds around.

The true story unfolded over the next day or so that this whole thing was orchestrated in support of the Yom Kippur War (1973 Arab-Israeli War) which took place between 6 and 26 October 1973. The surprise attack by both Egypt and Syria, of course, was begun on the Jewish holiday of Yom Kippur, and took place in the Sinai and Golan Heights. The Israelis were caught off guard the first day or so before they regained the advantage, so I'm guessing this story took place October 6 or 7. Knowing that now, the phrase about us being kept in reserve now meant something. Who knew how long that war would last and what direction it would take?

One thing I know - I'll never forget that night!

First Alert Tour

E.G. "Buck" Shuler, Jr.

Between October of 1960 and March of 1961, I went through advanced training as a B-52 pilot, with an end assignment to the 9th Bomb Squadron, 7th Bomb Wing at Carswell AFB, Texas. Advanced training in those days was an extended series of TDYs, which included nuclear weapons school at McConnell AFB, Kansas, academic ground school at Castle AFB, California, flying training in the B-52E at Walker AFB, New Mexico (where I was crewed up with my Carswell crew), and finally survival school at Stead AFB, Nevada.

I arrived at Carswell in late March of 1961 after getting married and was anxious to get started being a part of the nuclear deterrent behind the SAC shield. Following the completion of local unit requirements, meeting the Wing Commander, crew certification, etc., our Ready crew, under the command of Major Herman St. John and with me, a brand new first lieutenant as deputy aircraft commander, was scheduled for our first alert tour. The entire crew was excited about our important responsibility. We reported to the alert facility, cleared gate security, met with the crew we were relieving and proceeded to our B-52F, one of eight aircraft standing nuclear ground alert. At the aircraft we were admitted after passing the code number of the day to the security policeman on point guard. We completed a walk around of the aircraft while the off going crew removed their gear. We then placed our crew gear aboard, signed the required paperwork and formally relieved the off-going crew. Now this baby was ours and we proceeded back to the alert facility to receive the daily briefing.

In 1961 the 7th Bomb Wing scheduled seven day alert tours with change over on Tuesday mornings. The 9th Bomb Squadron, with 16 B-52F aircraft, was the only bomb squadron assigned to the wing at that time. Under the 50% alert concept, half of the assigned aircraft were on nuclear ground alert ready to launch within 15 minutes and proceed to strike assigned targets upon receipt of a properly authenticated message. This was the height of the "Cold War" and SAC was nearing its peak strength.

At about dusk dark on that Tuesday, the alert facility Klaxon sounded, which generated an immediate and electrifying response. Each crew was assigned an alert response vehicle, which in our case was an Air Force blue Ford station wagon. These vehicles were parked in a row adjacent to the alert facility. Traditionally the gunner was the designated driver and as we all raced to the vehicle our gunner, Airman Clyde Showers, assumed his position and started the engine. With a six man crew it was difficult to get everyone inside the vehicle so the tail gate was left down. Arriving a bit tardy I, along with our navigator Captain Maurice Miller, piled onto the extended tailgate. The drive to the aircraft was relatively short as the alert parking line was just across the ramp. For close in security there was a rope stretched on stanchions in front of the aircraft. Riding looking aft, as Clyde slowed to let the security policeman drop the rope, I assumed we had arrived and bailed out just as Clyde speeded back up going to the vehicle parking spot. I went ass over tea kettle, tore a hole in the knee of my flight suit, and skinning my knee on the concrete causing some degree of bleeding.

Dazed a bit, I recovered and charged up the hatch to my right seat position. Johnny St. John was already running the start engine check list while strapping in. I began starting the engines while also strapping in. The procedure was to start two engines using pyrotechnic cartridges and then advancing those two engines to about 85% to provide sufficient bleed air to start the remaining six engines. The cartridge on number eight engine fired properly, but in the process the engine torched, with flames coming out of the intake as well as the exhaust. Additionally, raw fuel was dumped on the ramp and the exhaust flames lit off the fuel on the ramp. Well at dusk dark, all of this fire lit up the entire ramp! The crew chief was scrambling to get a fire extinguisher focused on the problem. By then the command post controller, Major Tony Gekakis, had started the bomber roll call. He got to us and I responded, "Stand-by we have a fire going!", to which Tony said, "Roger, what are your intentions?" Simultaneously, Major St. John swung the closed fist of his right hand and hit me in the chest as hard as he could while yelling, "Don't say FIRE on the radio!" The crew chief was finally able to knock the fire down and we were able to complete our required actions. I learned several obvious lessons from this event.

Many years of alert duty and countless Klaxon responses followed, but that was a very inauspicious beginning for an officer who went on to serve as commander of two bomb wings, commander of two air divisions, SAC Director of Operations and Commander of Eighth Air Force. Many thanks for letting this old crew dog share some memories in the outstanding *"We Were Crewdogs"* series.

Morning Alert briefing.

The Day the Music Died
John R. Cate

The year was 1991 and although no B-52 gunners knew it, in less than a year we would be faced with life altering decisions and changes unknown since the days of Operation Arc Light, some 24 years prior. Just the year before the 60th Bomb Squadron, Andersen AFB, Guam had been deactivated and B-52s left the island for good, ending the bomber's continuous presence in the Pacific theater that had been ongoing since the Vietnam War. The handwriting was on the wall.

After leaving Guam, five gunners, including myself were reassigned to the 325th Bomb Squadron, Fairchild AFB, Washington. Looking back on it, I'm sure none of us knew what to fully expect. One thing was certain though; "pulling alert" would once again become a way of life for all of us.

To be sure, no matter which SAC base you were assigned to, once you were EWO certified, you would be pulling alert, that much hadn't changed. You knew you would pull alert at least once a month, sometimes more often but that rarely happened. You and your crew still went on alert on Thursday morning and you still got off alert the following Thursday morning. You knew you could expect to have at least one exercise per alert tour and if there hadn't been a "mover" for the month, you could expect one of those, too. On Monday mornings you still had the chance of a "no-notice" Emergency Procedures or EP test, brought to you by your local, friendly Stan Eval. No sweat though, just another way to serve as the example. After getting off alert, you still got four days of Combat Crew Rest and Relaxation or C^2R^2 or as it was more commonly known, C-Square. An alert pager was your best friend for the week. Daily trips to the BX with your crew were still the norm; you still had "reserved" back row seating at the base theater. For some, poker was still a pretty good way to pass the time.

Several aspects of pulling alert have changed over the years. For one, all B-52 gunners are now Command and Control Procedure (CCP) qualified. Most, if not all alert facilities have been remodeled over the past five years. While you still shared a room, now most rooms had cable television and the chow hall had three large windows. Maybe the view wasn't the best. After all, you could only see the runway hammerhead and Christmas Tree, but at least you could look outside while you dined.

Fairchild's alert facility has just been reopened after an extensive, year-long remodel. The chow hall had been completely removed, replaced with single rooms for the Flight Commander, Senior Pilot, Radar Nav, EW and Gunner and each of those rooms had cable television with a VCR, refrigerator, phone and semi-private bathroom. A separate building had been constructed just outside the entry control point to the alert facility and it had a modern, updated chow hall and a fully equipped gym with a weight room and two racketball courts. There were also several rooms set aside for family visitation. All-in-all, it was not a bad place to pull an alert tour.

The date was Thursday, September 26, 1991. I was on crew E-25, one of the six crews assigned to B Flight and B Flight assumed alert that day. It was the start of another alert tour. How many have there been? I wonder. Although I've been on flying status for 11 years at that point, I had only been a gunner for eight years. I completed gunner training at Castle AFB in June, 1980 and after three years with the 28th Bomb Squadron, Robins AFB, I was requested to retrain as an airborne radio operator. The B-52D model had just been pulled from the inventory and there was an overage of gunners. After three years of flying the National Emergency Airborne Command Post (NEACP), I made my way back to the BUFF.

I located my crew's sortie truck and after loading my bags, I met up with the other on-coming gunners for breakfast. After the morning briefing, all the on-coming crews headed out to their individual aircraft to sign for the aircraft, EWO box, tickets and weapons. As is the custom, I got a briefing from my sortie's off-going gunner. He told me my jet had only been on alert for a week and there had been an exercise (mover) the day before. The Fire Control System (FCS) checked out Code 1, no malfunctions. My crew preflighted the aircraft and each of us set up our individual crew position the way we liked it. After preflight, all the crews made their way to the Wing Intel Vault for CCP study. We were ahead of schedule, so the pilot decided to stop at the squadron for a few minutes. He had just upgraded to IP and needed to see the Ops Officer. I went by to see the Squadron Gunner, SMSgt Pete Karjanis. Pete and I had been friends since our days together with the 28th BMS. We talked for a few minutes and he mentioned there was a free taco bar at the NCO Club Friday night and most of the gunners were planning on being in attendance. I told him the alert gunners would be there as well. I spread the word at CCP study.

Friday, September 27, 1991. At 1700 hrs, I "commandeer" my crew's sortie truck and we (the alert gunners) head for the NCO Club. Actually it's no longer just an NCO Club, but is now a "combined" club. The Officers' Club is on the right side of the building and the NCO Club is on the left. Each club has its own casual lounge, formal bar and dining room, but shares a common kitchen. Upon arriving, we head for the casual lounge and find it crowded. The big screen television is on, but no one is paying attention. We spot Pete and MSgt Keith Krebs setting at the bar. Keith is our Training Flight Gunner and a good friend of mine. Keith lives two houses down from me in base housing. He knows the H-model gun system inside and out and got me up to speed and on instructor orders in record time when I PCS'd in from Andersen AFB. I get a plate full of tacos and grab a seat at the bar and talk with Pete and Keith. I look around and see several other friends of mine -friends from the maintenance squadrons, plus a few boom operators. Most of the gunners are there, too. There's good music on the stereo, guys setting around talking, catching up on the week, a few guys playing darts.

Suddenly, someone tells the bartender to turn off the stereo and change the television channel to CNN. A news commentator comes on and tells us to stand by for a special announcement from the President of the United States. As President Bush comes on the screen, the bar grows totally quiet. The President begins by saying "My fellow Americans..." As we listen, we are all astounded at what we hear. President Bush begins by telling the nation that the Soviet Union has been defeated and the Cold War has been won! The President next tells the nation that effective immediately all B-52s on nuclear alert as well as all land based inter-continental ballistic missiles (ICBM) will stand down. President Bush also states that beginning 0001 hrs, Eastern Standard Time (EST), Saturday September 28, 1991, all alert B-52s will have their nuclear weapons downloaded and be removed from their individual alert parking spots. He further states that this will be accomplished no later than 1200 hrs, EST in order for Soviet satellite imagery to verify these actions. I look at Pete, then Keith, and their facial expressions are the same as mine - one of shock and disbelief. Not surprisingly our alert pagers go off and Czar Control broadcasts, "This is Czar Control, Restricted Alert - All alert crews return to the alert facility immediately." The alert gunners head for our sortie truck and as we drive back to the alert facility, no one says a word.

At 1600 hrs all alert crews meet with the Wing Commander, Col Weinman. He tells us we are restricted to the alert facility until further notice and at the completion of his briefing we are to report to our individual aircraft and remove the EWO boxes and tickets, along with our flight gear. He further briefs us that we are to leave the alert facility at 0800 hrs, Saturday morning. At the conclusion of his briefing, he directs us not to contact our spouses or families. As we leave the alert facility and head for our aircraft, we can see a convoy of special weapons trailers and load crews coming down the taxiway toward the alert aircraft parking area, or Christmas Tree as it is known. Wing Intel goes to each aircraft and removes the EWO boxes and tickets. We remove our flight gear and load it into our sortie truck. As we leave a nuclear loaded B-52 for the last time, my Radar Nav steps into the right forward wheel well and "pops" open the bomb bay doors and my crew steps into the bomb bay. I reach out and run my hand across the nose of one of the nuclear weapons and we all turn and leave. No one looks back.

It's dark now and the Christmas Tree lights are on. All through the night flight crews stand outside the alert facility and watch as each

B-52 is downloaded and the weapons are convoyed back to the nuclear weapons storage area. Finally each aircraft is towed back to the flightline by OMS and parked. It's about 0300 hrs and the Christmas Tree sets empty and the "stadium" lights are turned out. No one goes to bed; it's like everyone wants to stay up and experience the last few hours of sitting nuclear alert. After all, this is a story we'll tell our grandchildren. The gunners all meet in my room and we talk through the rest of the night. We talk about what we think the future will hold, now that we're no longer pulling alert. We talk about the past, share a few war stories and for what may be the last time, just enjoy each other's company. My crew decides to meet that morning for breakfast at 0700 hrs.

After breakfast we load our sortie truck with our flight gear and personal bags and get in line with the other crews to leave the alert facility for the final time. We are all surprised to find both gates of the "Sally Port" open and drive through without stopping. By 0800 hrs all crews are gone and the alert facility stands empty. Later, that afternoon I step out my back door to have a beer and I instinctively look toward the alert facility. For the first time I don't see any vertical stabilizers.

On Tuesday, October 1, 1991 the Air Force announces that the gunner position has been eliminated from the B-52. On Friday, October 11, 1991 I have my final ride in a B-52. On the way back the pilot, Lt Col Whittenburg asks if there is anything "special" I want to do. I tell him I'd like to do a low approach over the Grand Coulee Dam. He contacts ATC and requests a low approach and five minute delay in route over the dam. Seattle Center approves his request and I move up to the IP seat and just enjoy the ride. After a couple of low approaches we head back to Fairchild AFB for a full stop.

Just like that my B-52 career is over. By Friday, October 15, 1991, I have an assignment to the 552 ACW, Tinker AFB, as an E-3 Radio Operator. I leave Fairchild AFB on Friday, November 1, 1991. As it turns out I'm the first gunner to arrive at Tinker AFB. Rory Koon is the second, arriving about three weeks later. Pete Karjanis leaves Fairchild AFB about six months later. He too has an assignment to Tinker AFB as an E-3 Radio Operator. Over the next year approximately 42 additional gunners, among them Cat Mason, Rory Alaniz, Ed James, Jeff Rowley, Joe McCollum and Jim Galambos, are assigned to the 552 ACW. The first thing we do is establish a gunners association. Bryan Luby and Tex Thompson suggests the name G.U.N.S. or Gunners Under New Specialties. It seemed appropriate.

Due to a massive Air Force reorganization, SAC and TAC merge and become the newly formed Air Combat Command (ACC). The date is June 1, 1992. Strategic Air Command, the most powerful air combat command the world has ever known, defender of our freedom, the command who brought the Soviet Union to her knees and won the Cold War, is dissolved with the stroke of a pen. Its 40 odd years of service and protection to our great nation, the 40 odd years of tradition, pride and sacrifice by thousands of men and women, are all placed in a box and put on the shelf. History records that the Army Air Corps trained 347,236 gunners during World War II. During my career as a B-52 gunner there were, on average, 450 gunners on flight status at any given time. Air Force Association President Michael M. Dunn offers these facts for our consideration: "When I entered the USAF in 1972, the average age of a USAF aircraft was about 8 years old. Today it is approaching 25 years old. The active duty Air Force presently has 333,000 people today. It is smaller than the Army Air Corps was on December 7, 1941. Air Force people have paid a heavy price for our security. Most people do not realize that in World War II the 8th Air Force alone lost more people (killed) than the entire U.S. Marine Corps lost in the war, and this is not to denigrate the contributions of the world's greatest Marine Corps."

Truman Smith, the author of *"The Wrong Stuff, The Adventures and Misadventures of an 8th Air Force Aviator"* writes "The danger is in not knowing. And in knowing, not remembering the way it was." C'est La Vie. (Thanks, Pete)

The Schedule
Doug Cooper

"He who controls the schedule controls the world." I don't know who first said that but it was one of the canons of the Strategic Air Command. At least it was when I arrived at Beale AFB back in 1968, a newly minted Electronic Warfare Officer straight out of Castle AFB Combat Crew Training.

The Scheduling Office was a large room filled with desks and field grade officers whose fingers were forever stained with grease pencil residue. Many were getting long in the tooth and were themselves all veterans of the crew force. The front wall of the room was covered with large sheets of plastic under which were thousands of lines and boxes. On the lines were the numbers of the crews and the names of individuals. Each box was a day on the calendar. Only after many months of scheduling did one actually understand the codes therein.

My first aircraft commander, Phil Doud, showed me the schedule and explained that, for all intents and purposes, it was set in concrete with only occasional changes being approved by senior staff. Occasionally, somebody would go DNIF (duty not involving flying) and a substitute would be put in; but, we had a lot of spares and so substitutions seldom reached into the crew pool. Acts of God, such as Inspector General (IG) visits, could drastically alter the schedule; however, only temporarily. After the IG departed, the schedule resumed with a vengeance since there was lots of time that needed to be made up.

We all learned to live with the schedule. You could look downstream up to three months to plan a trip to Lake Tahoe, a wedding (yours, maybe), or some other event. You knew when you would be on alert, flying or in some other type of training. This was your life story, written in advance for you by other people. In its own way, it had a certain pleasant rhythm. Life was organized. You always knew what to do.

1968 turned into 1969. 1969 turned into 1970. The schedule began a slow and painful evolution from strict organization to "who can we screw with this requirement." "Whose turn is it in the barrel?" We cried, "Give us back the old schedule. We're sorry that we ever complained. How can things be any worse?"

Chaos reigned. You didn't know from one week to the next if you were going to fly or pull alert. Everyone was either getting ready to depart for Guam, on Guam, or returning to Guam. No one was getting credit for a Southeast Asian tour because each trip only lasted 189 days - which, I believe, was the minimum for tour credit. Guys were bailing out of the service like lemmings going over the cliff. The D model bases were beefing up their crew strengths. I PCSd to Carswell around this time.

Then came the fall of 1972. All of us went to Guam - all of us. It was standing room only at the club and quarters were so sparse that many of us were billeted in the beach hotels (hardship).

I learned to never challenge worse. I thought I had experienced it all until Christmas of 1972 when the crews were housed in tents, next to the engine test stands at Andersen.

The moral of this story: never complain about the schedule. It can get a helluva lot worse.

Jaws Bites CCTS
Rich Vande Vorde

I enjoyed being a Navigator on the B-52, but the normal crew progression was not to stay a Navigator but instead to upgrade to Radar Navigator (RN). The Squadron decided I should upgrade within the Squadron instead of the proven method of attending Combat Crew Training School (CCTS). So, needless to say, the Wing schedulers and Training Flight did not have a detailed plan for which instructors would fly with me to keep my training as coherent as needed for a timely completion.

Well, after about two months and only three missions completed in my RN training, the Squadron changed their minds and decided I really should attend CCTS. There was a short orientation to get organized. We met the instructor and our student crews. They also informed us that our evaluation at the end of our training was going to be unusual. Our check ride was to be performed by a Headquarter Standardization Evaluation Crew from Barksdale, AFB whenever we completed our training.

With that, we started planning for our first CCTS mission. We arrived at the training facility ready to do our thing when we heard the KLAXON. To our surprise, the Operation Readiness Inspection (ORI) team had arrived to give the wing an inspection. Our first flight in CCTS was cancelled. Jaws had arrived. Jaws was the nickname given to the ORI low level route that had the largest percentage of bad scores in history, I believe.

Since so many wings had failed their ORI flying missions through that route, CCTS crews were designated to fly the route themselves to prove that everything was realistic, I suppose. I don't know exactly what they did to plan for their missions; however, they completed it successfully and wings had to fly the same missions for their remake. That could be another whole story in itself!

So...back to CCTS training. We continued the next few weeks completing the requirements for upgrading. Late in the training, the instructor asked me what types of bomb runs I wanted to do. I wasn't sure but I finally decided to do some tricky types of runs. I wanted to practice ones to help not only be a good RN, but to be the best RN. For example, we did one without a heading system, i.e. heading up instead of north-oriented and quadruple Bombing and Navigation System (BNS) synchronize releases less than two minutes apart. We had a few scores that were outside the acceptable range, but CCTS is the place to practice. So I did not worry about whether the people at the Wing Mission Review meeting saw them.

We also had many T-10 simulator periods where I asked the instructor to initiate malfunctions with the BNS. Then we could learn procedures to recover from them and still have a reliable release. Many of the problems could be detected by making a series of crosschecks including checking the manual timing with the BNS. During a single two-hour T-10 trainer we could perform about eight or nine bomb runs in order to perfect our techniques. On the later bomb runs during the T-10 period we would do all the crosschecks, but the instructor might not give a malfunction at all. That trained us to be ready for almost any type of bomb run malfunction.

After all the training was complete, we were nearly ready for our evaluation. The flight just before our check ride with the evaluation crew, we flew with a different CCTS instructor crew. The Wing wanted to make sure we were ready for the Barksdale evaluators. The mission profile was exactly like our checkride would be. Mission planning through the flight to mission debrief went like clock work. We were ready.

Two days later we met the headquarters evaluators and planned an identical mission. As we started mission planning the KLAXON sounded. Yes, the ORI team was back for the remake of Jaws. The Barksdale evaluation crew went back home to wait for us to fly again. It seemed strange that our first and last CCTS missions were cancelled because of an ORI (Jaws). After the wing got back to normal operations, the CCTS leaders requested that the evaluators fly with a different crew, with no success, they still required us to fly with them.

The Barksdale crew came back. The mission was very smooth, except for one major area, the low level bomb run. We went to

Hastings low level. I did the standard evaluation bomb run, a BNS synchronous release followed by an alternate release, time and heading by the Navigator. We performed all the crosschecks we had learned during flights and T-10s. Then the site called back our scores, the first release was reliable, but to my surprise the alternate release was not. Did another Jaws bite us?

As it turned out, the Nav had used the speed at the start of our popup and pushover maneuver instead of the average speed. When he calculated the time to release, it was too long. However, we both were still qualified from the Barksdale evaluator's perspective. We had performed the procedures correctly and the small time error was not enough for anyone to detect. These Jaws did not bite us! However, the Senior Wing Standardization Evaluation Crew did not like the fact that the Nav was still qualified despite the inaccurate calculation and made it perfectly clear to our crew and the squadron that the performance was unsatisfactory.

Despite that, the Squadron was ready for us to get back and we spent the next week or so preparing to certify to pull Alert the week after that. That was when I was assigned to the Best B-52 Crew in SAC, which eventually became S-09. (Refer to "We Were Crew Dogs II", page 102).

That experience may be similar to many aircrew members' experiences from Carswell AFB during Jaws. But one thing is sure, being a B-52 aircrew member is special and CCTS did a great job.

Engine Fire
Lothar Deil

As I look back on my time flying the G-model BUFF at Fairchild AFB, Washington, one unique incident stands out as one of the most memorable I encountered while on a crew. I had been on station for almost four years. The command was just recovering from the fuel crisis and an abundance of copilots that plagued us during the late 70's. Alley Oop Squadron was so flush with copilots that they took to designating them as "Junior" and "Senior" and I had recently been promoted to "Senior" status. Basically, it meant that they figured I was so close to upgrade that they needed to give me more than one touch and go per sortie. I was the copilot on Crew R-21, and our crew was scheduled for a routine post-alert 8.5-hour B-52G training sortie. The best part was that there would not be another copilot on this flight, which meant more flying time for me. We were still adjusting to our new pilot, a real change from Capt Capitosti. Capt Capitosti was considerably more animated and emotional than this new pilot (we had no idea how much the real contrast in personalities would be until after this flight). Our routine flight was scheduled to have all the usual events: Air Refueling, low level flying at OB 300, and then back to the pattern for a few touch-and-go's. Our call sign would be Buddy 23.

Of course, a "white-throated nit-picker" from Stan Eval was waiting for us at Base Ops, and our routine training flight now became a Check Ride for the pilots (which explained why a second copilot was not scheduled). I should have known.

The flight went quite well, even though our new pilot had only one flight with us. The rest of the crew had been together for almost three years and easily compensated for minor deviations. So, seven hours later, at 3,500 feet, we were inbound for the first touch-and-go off a straight-in approach to Fairchild AFB, crossing the approach to Geiger Field (Spokane International). ATC (Air Traffic Control) called out crossing traffic below us, a 737 going into Geiger Field. The cockpit lit up with what we initially though was the landing lights of the 737, except it was coming from the wrong side.

Our new pilot looked out the window, and with surprising calm, he said "Huh … It looks like the number three is on fire," which was alarming enough; however, he then started to give us a running commentary on the fire, "It's pretty cool, kind of orange and all." The lack of alarm in his voice was just surreal. I was sitting there thinking to myself, "Isn't this where we do the bold face?" I don't know if we were just tired, or if the pilot's calm attitude just set the tone for the rest of the crew, but I found myself stating, almost as a suggestion, "We should shut number three down, then."

He concurred calmly with, "That would be good; let's shut it down." He shut number three down, and the fire, although it didn't go completely out, seemed less intense. "We should pull the T handle, as well," I heard myself say matter-of-factly. Again, he said what seemed strangely just a little off, "That's excellent. Pulling number three."

ATC switched us to Tower, and I called in, "Buddy 23, declaring an emergency. We have a fire on number three. It appears out. This will be a full stop. There are seven onboard." My tone must have been really calm, because tower called back, "Buddy 23, are you really declaring an emergency?" So, I confirmed the emergency declaration, and Tower said that the fire trucks were on their way.

We completed the approach and landing checklist. The rest was a straight-forward full-stop landing, and as we rolled out, the pilot said "Now, we're all going to get out after we stop and meet in front of the airplane, OK?" So that's how it went. After the fire crew formed on number three and made sure the fire had burned out, we examined the engine. Apparently, one of the ejectors broke, and as it was melting, it bent and acted like a blow torch etching a large "L" into the cowling.

Our crew's flight debrief normally took about an hour to finish the paperwork and flightlog for scheduling, discussing what went well, what could have been better, etc. A check ride, however, would be 30 minutes to an hour (depending on how well or bad the ride was) for the Check Pilot to debrief the crew and then the pilots on how they did. For this particular flight, however, the entire time from landing to completion of the flight debrief took three solid hours. This was because before we could do any of the regular stuff we spent an hour answering questions from Command Post from what seemed like half of Strategic Air Command. Most of the questions were easy to answer (e.g. "When did the fire break out?", or "Who was flying when it happened?") Some were obtuse, like my favorite, "Why didn't the fire go out right away?"

So it was almost 10 PM when I drove home, where my wife, Carol, was wondering if I knew anything about the bomber on the news that landed tonight and set all the brush fires in Airway Heights.

"Yea," I replied matter-of-factly, "that was us."

Grand Teton Mountains

It's a Small World After All
Bill Dettmer

In 1974, I was a B-52 copilot on a Stan/Eval crew at Beale AFB in Northern California. Part of our training routine was to fly low-level navigation and bombing practice missions throughout the western United States. We never actually dropped any bombs during these practice missions, we just went through the procedures - over and over again, ad infinitum, year in and year out, winter and summer, rain or shine.

On one mission in August 1974, our crew was enroute to the Midwest to fly a low-level training mission in Nebraska. But as frequently happens during the summer, a line of thunderstorms developed over the Rocky Mountains east of Utah. The thunderstorms were more severe than usual, and the line stretched from as far as we could see (at least two hundred miles) to the north all the way out of sight to the south. The tops of these thunderstorms were much higher than we could climb—perhaps 45,000 to 50,000 feet. And the separation between individual cells (storms) was less than 15 miles. Given that it's mortally dangerous to fly through a big thunderstorm, and we had what appeared to be a solid wall of them ahead, we aborted our planned mission and turned north toward the Great Salt Lake at an altitude of about 31,000 feet.

We orbited for 20 minutes while we made radio contact with our wing headquarters back at Beale AFB and asked for some alternatives. The headquarters came back with a block of reserved times on a visual low level route that ran from southeastern Idaho westward and southward to Pyramid Lake, Nevada, just north of Reno. We "owned" that route for a solid hour and a half, meaning no other military airplane could use it during that time.

But at that point, our airplane was too heavy with fuel to descend and fly at low level, where turbulence often put heavy stresses on the airframe. We had to orbit for an hour to burn down fuel. Since the entry point for the low level route we were to fly was just over Driggs, Idaho, we decided to kill our time around there. Driggs is just west of the Grand Teton Mountains, which themselves are on the border between Wyoming and Idaho. We descended from 30,000 feet down to 18,000 (jets consume fuel faster at lower altitudes). As we got lighter and lighter, we could safely descend lower.

We still had about 20 minutes until our low level route entry time. I suggested we go on a little sightseeing tour over the Grand Tetons, visible just a few miles to our east. We canceled our instrument flight clearance, descended to 13,000 feet, and proceeded east under visual flight rules.

The profile of the Grand Tetons makes them the most recognized mountains in America. Grand Teton Peak itself is 13,766 feet in elevation. Mount Owen, the peak just to its north is 12,800 feet. The U-shaped pass between the two peaks is called "The Gunsight" because of the small V-shaped notch in the center of the pass. From the east, these mountains rise 8,000 feet up from the floor of Jackson Hole, but from the west the slope upward is much more gradual. It was an exciting effect when we cleared the Gunsight Pass, because we were flying 700 feet lower than the Grand Teton Peak and only 200 feet higher than Mount Owen. But it was a little disconcerting looking up at the big mountain as we went by at 300 knots. And then there was the instant of vertigo as we went from about 400 feet altitude above the ground to 8,400 feet. We were less than 1,000 feet laterally from both peaks.

After passing the Teton Range, we turned northward and flew at 13,000 feet up over Yellowstone National Park, circled Yellowstone Lake once, and headed back toward Driggs, Idaho, for our low level navigation run. Our "sightseeing tour" was over for the day.

Now fast-forward exactly 28 years to the month. In August 2002, I was driving across the U.S. from Seattle to Washington, D.C., with my oldest daughter, Melissa. I planned a day-and-a-half stay in Jackson Hole, where we would explore Grand Teton National Park and float down the Snake River in Jackson Hole on a raft.

Early one morning, we boarded a big rubber raft with six other people and a raft captain/guide, a young man in his mid-twenties. I pointed to the Gunsight Pass directly behind the raft captain, and I said to my daughter, "Do you see that pass just to the right of Grand Teton Peak? In August 1974, I flew a B-52 right through there, just lower than the big peak and just above the shorter one."

Before my daughter could respond, the raft captain looked at me and exclaimed, "That was you?!" I responded, "Yeah, but what could you possibly know about it? You weren't even born then."

"No," he replied, "but all the river guides still talk about that! Man, are they gonna be surprised when I tell them I met you!"

So I guess I'm notorious in Jackson Hole…at least in certain social circles!

"…And Now, the Rest of the Story!"

I thought about separating this from the account above, but since it happened about an hour after we overflew the Grand Tetons in 1974, I thought I'd just append it here.

After departing Yellowstone Park in our B-52, as described above, we flew southwest over Driggs, Idaho, and entered the low level navigation route, March 15-9, which followed an imaginary line (well, we had it plotted on our maps) over southern Idaho and into northern Nevada. To say the land out there is desolate is sugar-coating it. It's all stark volcanic mountains without a single tree, interrupted by dry lake beds and desert basins where little grows except sage and other desert plants. You can see for 40 miles, even at low altitude - and we flew very low.

In those days, to survive in wartime we would have had to fly below enemy radar coverage, whether in daytime or at night. That meant we had to be down almost "in the weeds" - less than 500 feet above the ground. Now, anyone who has flown in a light airplane knows what it's like to fly that low. And even if you've only flown in an airliner, think about looking out the window after takeoff as the plane climbs out. You're at about 400 to 500 feet when you reach the departure end of the runway. That's how low we flew. But when you took off in your airliner, you were flying at about 200 miles per hour at that point. We typically flew between 380 and 400 miles per hour. The ground appears to move very rapidly below the airplane at that speed and altitude. In fact, flying like that is a hell of an adrenaline rush.

So, there we were, flying down a mountain valley in northern Nevada at 380 miles per hour, some 400 feet above the ground on a hot August afternoon. You could see sagebrush and the occasional coyote, antelope, or deer running like hell to get away from us. (Besides covering a lot of sky from their perspective, we made a lot of noise.)

As I looked ahead, I could see a rooster-tail of dust flying up from the desert floor about 10 miles ahead. A local rancher in his pick-up truck, I thought. And I smiled at the thought that we were going to scare the bejesus out of him when we crossed over his truck. We used to refer to that as letting the local citizens hear "the sound of freedom."

I kept looking ahead at the dust cloud, and I noticed a funny thing - there was no road where it was located. It was just desert sagebrush out there. It didn't seem right that somebody would be driving fast enough off-road out there to generate such a dust cloud.

We got to within two miles, and I got a better look at what was raising the dust. It wasn't a vehicle - it was a herd of about 40 wild mustang horses. They were galloping across our projected track from right to left. When we were about a mile from them, they crossed our nose. Ten seconds later they were just outside my left window, 400 feet below, and turning to get away from us. I announced to the crew on the interphone that we had wild mustangs outside. The four crewmembers in the back, who had no windows, wanted a look at them. One by one, they came forward. We put the bomber into a shallow, 15-degree bank left turn and slowed down to about 200 miles per hour.

This kept the mustangs in the center of the circle we were flying around them, while we flew the circumference. And because the left wing was low, the horses were easily visible out the window. It takes about four minutes to complete a 360-degree turn at that airspeed and bank angle, more than enough time for everybody who wanted to see the horses to come have a look.

While we were turning around the horses, I noticed something unusual. They, too, were turning left in a desperate attempt to get away from us. As we kept turning, they kept turning, until eventually we completed our 360-degree turn and continued on our original course. By this time, they too had completed a 360-degree turn and continued in their original direction.

It wasn't until later in the flight that it occurred to me that we had actually herded wild mustangs in a B-52! ("Giddyup! Yee-hah!") To this day, I believe we are the only pilots in the history of the world ever to herd wild mustangs in a B-52.

She's a Tuff Old Bird

Ben Barnard

Flying low level was one of the most challenging parts of training in the B-52....but, it was fun. Getting down to 400-500 feet above ground level (AGL) in an airplane that large and flying fairly close to "redline" airspeed is rather invigorating for even the best aviators. The particular low level I am going to describe had some additional excitement - excitement we didn't even know about until after we had landed.

It started out as a routine mission - even though it was a checkride (more formally called a SACR 60-4) for an initial instructor pilot evaluation for the pilot. I was the evaluator. The mission profile called for air refueling, low level navigation and bombing, and back to the pattern for several approaches, a touch-and-go, and a full stop.

Takeoff and climb out were normal as was the air refueling. Post A/R was uneventful as we prepared to enter the Hastings, Nebraska low level route. It was a beautiful spring day - CAVU—clear and visibility unlimited. And, as opposed to some days during the summer, it was unusually smooth as we began the route. Everything was progressing nicely, and with the terrain avoidance check completed we began the ingress to the target area. I was in the jump seat between the pilots doing normal evaluating duties, when I noticed a flock of birds at 12 o'clock dead on at our altitude. Some of them had already starting taking evasive action (birds tuck and dive if they sense they are about to be run over by a large alien). Since I didn't have much reaction time, I keyed my mike and said, "Birds, 12 o'clock." Then, I immediately put my head down on the center console, because I thought they were sure to come through the windscreen.

I heard no crashing sound and felt nothing either. Usually when you take a bird strike, you can feel it in the cockpit. As I resumed my normal upright position, I immediately looked at the engine instruments to ensure we still had eight normally operating motors. All looked normal. I, then, asked the pilots if they had seen the birds and they both said they had not. I told them to check their respective wings to see if there were any indications of a strike. Both reported that all appeared normal, so I assumed we had gotten through the flock without taking a hit. In addition to all of the engines operating normally, we had no hydraulic or electric problems, which can occur if you take on a bird.

As such, we continued with what we had planned. We finished flying the route, accomplishing two bomb runs as I recall, performed a normal climb back to altitude, and headed back to home station in order to perform traffic pattern transition work.

The instructor pilot candidate and the copilot were in the seats for the flaps-up approach and go around as well as the simulated six engine approach and go. I got into the seat for the touch-and-go since the instructor pilot candidate was still technically "unqualified." We accomplished the touch-and-go with simulated loss of outboard engine on the go part, and then we pulled up into the visual pattern for the full stop landing. All, again, went as planned. We did the full stop, cleared the active runway, and began our taxi back to parking.

As normal, when we pulled onto the parking ramp, a crew chief was there to marshal us to the area where we would stop and shut down so that the Uke could push the aircraft back into the parking space.

After we stopped and the crew chief plugged his mike into the aircraft so we could communicate with him, he excitedly asked us if we had hit some birds. I responded by saying that I had seen some birds during low level, but I didn't think we had hit any. He then said, "Yes sir, you hit some birds." Notice he said birds - plural.

After we deplaned, some of the damage was obvious, but some was not. There was a 12 to 15 inch hole outboard of the number four nacelle and another hole in the number three engine cowling - neither of these was visible from the cockpit. We later found out that at least one bird entered the scoop at the bottom of the vertical stabilizer and tore up all kinds of stuff in the vertical fin. The airplane was grounded for the better part of two weeks while the maintainers, in particular, the sheet metal shop put things back together. They had to lay the vertical fin parallel with the wings in order to fix the damage in that area. In all, they decided that we had taken at least three strikes, but suggested we had encountered four or five birds - and these weren't small birds, they were sand hill cranes - popular in that part of Nebraska at that time of year.

All of this happened and we had no indication in the cockpit, that we had taken that much damage. There were no abnormal indications on any of the instrumentation in the cockpit and no abnormal handling characteristics - quite a testament to the aircraft that we potentially would have to take into combat.

From that experience, I gained a lot of confidence about the amount of "battle" damage the airplane could sustain, and still keep flying and flying normally.

Crew compartment (right side of picture.)

B-52 Crash Survivor
Gregory Smith

It seems like many flying stories open with something like, "The flight began as a normal night mission; little did we know what was in store for us in only a few hours." Still, this is the short saga of one such typical nine-hour K.I. Sawyer AFB B-52 flight that ended up being the final flight of the aircraft nicknamed the Black Widow (60-0040) and some of its crew.

This December sortie consisted of an early evening takeoff, night air refueling, two hours of flying mountainous low level in the STRC (the series of low level routes and electronic bomb "plots" that is officially called the Strategic Training Range Complex), an hour and a half droning home, and another hour or so of beating up the pattern with instrument approaches and touch and go's. Before I dig into the specifics of that B-52 flying adventure, let me take you back a bit, and then I will bring this story back to the ground.

Having only recently graduated from initial B-52 training at the Combat Crew Training Squadron (CCTS) at Castle AFB, I was new to K.I. Sawyer. The program was a fun challenge, and it went very well; therefore, I felt ready for the tests that would come with all the new things involved in a fresh assignment. Even though there was almost

114

nothing new about the B-52 or the Strategic Air Command, it was an original and new world for me, and I was ready to conquer it. Since I had no experience with the military prior to pilot training, this was all novel to me.

After arriving at my new base in the thick woods of Michigan's Upper Peninsula, I was quickly in-processed into the 644th Bombardment Squadron. Training Flight took over the responsibility of my mission qualification, and I was introduced to a legend - Steven F. Nunn. I won't repeat any of the things that people thought the "F" stood for, but suffice it to say, he really knew the B-52. After accumulating his several thousand hours in the BUFF over the course of a couple of decades, his thorough knowledge could strike fear in the heart of a young copilot. He proceeded to grill, badger, and belittle the new copilots, stating that a little pain while at zero airspeed would help us remember during the pressure of a real emergency procedure (EP).

My training at Castle was in the B-52G, and since K.I. Sawyer flew "H's," I had to go through difference training. I didn't miss the water "lotto game" on takeoff we had with the G-model. Although the water injected system was reasonably reliable, you never knew when it wouldn't work, and it always seemed to increase the risks for our many heavy weight launches. We were told too many nasty stories of B-52 and KC-135 crashes that occurred on takeoff due to water problems. Some of the local tanker pilots wore a patch on their flight suit which seemed appropriate. It went something like: "KC-135A Steam Jet, built when they thought water would burn." One other nice thing about the H-model was the intercom panel. It seems like such a basic thing now, but back then it was one of the last vestiges of WWII-era Boeing airplanes besides the autopilot. It was part of why they called the H-model, the "Cadillac" of all things; something that seems a bit far fetched to some of today's younger generation of pilots. By the way, at Whiteman in the B-2 community these young folks are called SNAPS, an acronym for "Sensitive New Age Pilots." I guess that makes me officially old because I didn't make that cut. Having a bald cranium and no desire for the legal, but mind-altering in-flight drugs excluded me from that crowd. That's just fine with me.

Nevertheless, the "new" intercom allowed us to only listen to the radio or radios we wanted to hear and even allowed us a "private" channel for more exclusive conversations. It helped alleviate situations like I had created on a flight at Castle flying a G-model where I called

Flight Service asking for some weather and updated altimeter settings for a low level route only to be unknowingly answered by the instructor navigator downstairs. The crew got a big laugh out of it because I suckered for it hook, line, and sinker. The new intercom panel still required the use of a wafer switch which could be inadvertently left in the wrong position. The result was occasionally an embarrassing moment when, for instance, the mouth of a gunner would engage with some disparaging and vulgar counter-comment about the Wing Commander before realizing that his words had just left the jet over the command post's frequency. The mistake might have evaporated into the electronic ether except for the "poor" timing of the Wing Commander's daily romp on the ramp. His white-topped car and the radio carefully tuned to the command's common frequency provided him with more insight than necessary into just what some aircrew member thought of him.

Having completed my aircraft difference training, which emphasized the "advanced" nature of the newer model of the B-52, I was cut free of Training Flight and sent over to the large concrete building we called "the vault" that was across the street where I would begin my foray into the true mission of this awesome airplane. Here I was introduced to the salt mine, back-end of mission support and the vault creatures that carried out this thankless challenge.

This involved about three weeks spent in the vault pouring over the "Emergency War Order (EWO)" volumes in preparation to sit alert, ready to execute our EWO tasking (or, as we used to say, we were prepared to defend our northern border from the onslaught of invading Canadians - or something like that). Much of the study was just about the most boring information I'd ever read. The trouble was that buried in the General Lemay-era regulations were several things that could get you in some serious trouble. Whether it scripted detailed Klaxon procedures, described how to decode a message, or told you what you could wear on alert, some of it was also about the driest, yet most critical material I'd seen in my short experience with military guidance. I even had the opportunity to sign what, at the time, seemed complete reality to me. It was a memo for my crew requiring that in the event of a bailout during "nu-cu-lar war, toe-to-toe with the Ruskies..." where food would be scarce, I had to be willing to allow my body to be used as food in case I didn't survive. Fortunately, I caught on after watching a couple other new copilots sign the document and seeing their crews snickering off in a corner. Following an artificially pressure-filled mission brief to the Wing Commander where he signed me off as certified to carry out the mission, I was assigned to a crew and promptly put on alert. Now, alert was a complete culture of its own. I had heard wild stories of crews' TDY escapades, but what happened on alert seemed to be part of a whole different life. It was closer to a cross between living in college dorms and an elementary school sleepover. Antics abounded. Grown men behaved like children while simultaneously risking their lives protecting our country.

Every week, crews would gather in the snow filled parking lot outside the "sally port" next to the alert facility with their car or truck load of gear which included piles of suitcases, crew pubs bags, books, magazines, small weapons caches, and even rudimentary desktop computers, since laptops weren't yet popular or affordable. We'd pile into a bus and proceed to get inspected by our weekly prison-keepers, the security police, as we trudged our way to the alert facility and to what would be our "Alcatraz" for the next week.

After dumping our bags in our "luxury suites for two," we'd hurry out to the airplanes in the alert stubs for the official changeover. What followed was a ritual lived out by many BUFF crews over the decades. Special care was taken to inventory the command and control documents, refill the small cockpit with a truckload of unnecessary gear, perform an aircraft "walk-around" as if it wasn't all there, and of course, inspect the weapons. That occasionally provided a "Slim Pickens moment" as someone might demonstrate how he rode the bomb out of the B-52 bomb bay in the movie, Dr Strangelove. Finally, if we weren't totally frozen by then, we would actually hurry through the daily alert preflight checklist and confirm that some of the basics were still working. If there was a delay, it might be necessary to grab the spotlight which was plugged in right behind the copilot's seat and shine it on your boots at extremely close range until the polish began to smoke. The light beam was an effective way to turn some of the noise from the soot-belching generator into useful heat (just one more useful lifesaving tool designed into the great B-52). This technique usually allowed your toes to weather the temperatures until you could be rescued by the warmth of the 1950s era alert facility.

At last the morning outdoor work was done, and you could look forward to a greasy 25 cent omelet, a monstrous cinnamon roll and a hot cup of acidic coffee. Although it didn't matter to me at the time, the coffee seemed so strong that it reminded me of the liquid sulfur that

I used in my industrial materials lab back in college. Whatever was in it, the black substance in the Styrofoam cup kept me awake through most of the day's activities and still didn't interfere with occasional afternoon Nuclear Alert Procedures (more commonly known as an N-A-P).

With breakfast complete, it was off to the required EWO study at the vault. We first were required to sign out with the controller's office where one of the full time guys was Welby, MD. Yes, that was his real name. According to him, he had liked the 70's TV show, "Marcus Welby, MD" so much that he legally changed his name. Now he was known as Sgt, MD. Like robots, we all paraded to the "six-pack" for the ride to the vault.

EWO study began by signing out sortie bags and familiarizing ourselves with our mission. As a new guy, I plodded through the documents and publications, not completely bored yet with the material. The monotony was usually broken by doing things like playing lottery with the serial numbers on dollar bills, where the loser had to buy sodas. In later years we would play SIOP (single integrated operational order) Trivia where the instructor would toss candy to the person with the correct answer. That could even be mixed in with movie trivia to pass the time, but we still had to fill in the open book test question answers (not too hard for a whole crew, some of whom actually had significant experience). Time seemed to be the key factor in bag study. We seemed to need a certain couple of hours absorbing the material. Thus, whether it was spent in careful reading, or soaking the information up with your cranium flattened into the fold of a book binding, as long as your drool didn't get the pages too wet, you'd studied enough.

And so the days passed by, filled with morning briefings, daily alert preflights, additional antics in the bedrooms and hallways, "financial committee meetings" (poker) in the game room, incessant eating, and regular trips to the base exchange to basically confirm that virtually nothing had moved from the shelves since the last time we'd looked. Pranks were standard especially for virgin alert guys like me. Beds were short-sheeted, baby powder was put on the tops of ceiling fans, and beds were moved into the snow bank. In later years, we even pulled a guy out of bed, poured honey all over his nearly naked body, and broke open a feather pillow, pouring its contents all over him in a modern day tarring and feathering. "Start carts" were another favorite where one's wrists, feet and neck openings were sealed with speed tape and a shaving cream can was popped open with a can opener in the neck zipper opening of the flight suit. This allowed the contents of the shaving cream to be emptied quickly into the flight suit. A can or two produced a nice Michelin Man look.

The days moved right along, and before we knew it, our week was over and we could embark on a three and a half day long weekend filled with "honey do" projects and general homework. Soon it was on to the next week's "normal" schedule back in the squadron. One surprise I had waiting for me was a crew change. It turned out that the Director of Operations felt that I would be unduly negatively influenced by the attitudes of some of my fellow crewmembers, and he moved me to another crew. It seemed fine with me since I didn't really know anybody anyway and also because they were scheduled to fly that week. That would allow me a chance to get back in the air right away.

After an uneventful introduction to the crew after our morning weather briefing, we set our hearts on the task at hand. It doesn't seem possible now, but we managed to cram about three hours of mission planning into a whole day. Sure, there were several important things to talk about, but eight hours or so of mulling over our flight was more than adequate for all but the most challenging missions. Many critical things had to take place, like giving your lunch orders to the gunner so he could call them in to the In Flight Kitchen. At least your lunch was one surprise on the flight that usually couldn't kill you, but it could leave some temporary damage. Also, we had to wait for the poor navigator to do what now seems like indentured servant labor as he manually drew out a chart in order to get the timing just right. Then he would again manually figure out all the radials and distances for our FAA flight plan. Today we do that in seconds on a computer. Of course no mission planning day would be the same without the copilot sweating over the weight and balance sheet with one of the infamous slide rule-like slip sticks. This general busyness culminated in a crew briefing which could take over an hour sometimes. In our small, self-help-built rooms with no ventilation, it provided simultaneous preparation for surviving the B-52 cockpit on the ramp in the middle of the summer when no engines were running. Finally, for us young, wet behind the ears tykes, we had to recite the mantra, "I'll take off and land, strapped in position and remain in position until the aircraft comes to a complete stop…" There's more, but I'll spare you any painful memories that may dredge up.

As mentioned earlier, the sortie launch time was after dark, which was actually late afternoon since it was winter and the Upper Peninsula of Michigan remained on eastern time even though geographically it should be on central time. Generally, everything on the flight was uneventful. Strong low-level winds kept us at Instrument Flight Rules (IFR) altitudes for a short while but subsided, enabling us to descend to terrain avoidance (TA) altitudes before reaching the first bomb run. Clear weather dominated the entire flight, and the stars were shining brightly across the whole Midwest with an occasional hint of the Northern Lights off toward Canada visible on the way home.

We returned to the local radar pattern where I flew the first approach and the aircraft commander flew the second. By that time it was one o'clock in the morning and the crew was fairly quiet. On the downwind leg of the third pattern, I climbed out of the right seat and sat down at the tenth man position (facing backwards on the floor behind the Instructor Pilot's seat). The other copilot climbed into my vacated seat and flew what would be that airframe's last touch and go. The approach was steady, and the landing superb. In order to take a mental picture of the after-touchdown landing attitude, I unstrapped my seat belt and looked up over the IP's shoulder while we rolled on the runway. As the BUFF lifted back into the air, I sat back into my seat. Not in any hurry to strap myself back to the pile of cushions, I paused momentarily to reflect on how I could improve my landings, neglecting to immediately fasten my seatbelt. We accelerated through 150 knots and climbed toward 50 feet.

Seconds later, the airplane shuddered, pitching downward as an explosion erupted in the aft fuselage. Instantaneous thoughts of disbelief and confusion sped through my mind as we slammed into the ground with an estimated 23 Gs. In another instant we lay motionless with gear and equipment strewn about, only visible by the blaze which roared through the rear bulkhead. Time compression kicked in and, although I didn't realize it at the time, my senses were keenly aware as so many things happened so quickly. It seems like an hour now, although in reality it was only a few minutes. As the smoke billowed from the growing flames above the Electronic Warfare Officer (EW) and Gunner's panel, I began to realize that this was more than just a blown tire. The EW and Gunner's whole wall of equipment had been pushed in on them, and a gaping hole was ripped open above them with flames leaping into the crew cabin. I was still leaning against the tenth man hammock while pubs cases and divert kit bags covered me. The communications kit (which was lying unsecured in the bunk) had also hit me in the head. The impact broke my mask and helmet, slicing my chin and fracturing a bone in my ear canal. Thinking the cockpit was still attached to the airplane, I realized that the forward body fuel tank could explode at any moment. My legs didn't feel like they would work after all the hard hits from the unrestrained equipment. I released myself from the parachute and crawled up to the IP seat. I couldn't see the pilots very well as thick smoke filled the cockpit. The only light was that from the nearby alert facility only about a hundred yards off the nose. Right next to it were six fully loaded bombers and six fully loaded tankers, ready for war and daring us to get in their way—and we were. Shouting to the pilots, I asked them if they were okay and if they could get their windows open. I received negatives to both questions.

Barely able to breathe now from the suffocating, toxic vapors, I turned back toward the defense compartment where the flames continued licking at the insulation along the wall. I pulled my T-shirt over my nose and mouth to filter the air and began to take stock of the situation and the other crew members. The gunner and I beat the flames with our boots, hands and other equipment we could pick up. This was especially tough for him since one of his legs was broken. I turned to the EW who was screaming with pain as the entire EW suite of equipment was crushed in on his legs, resulting in a broken femur and other minor cuts and bruises. Next I shouted down into the blackness of the lower level navigator crew compartment. The Radar Navigator (RN) was the first to answer and was in relatively good spirits, considering all the mayhem, besides not knowing the status of his live ejection seat and having a serious compound fracture on his lower right leg. The navigator also sounded like he could survive despite having two broken ankles. I helped the RN on the ladder to the upper level; somehow, we knew that trying the crew entry hatch would be fruitless, and then moved back toward the gaping hole above the EW station. At that point, getting everyone out of the wreck seemed paramount, and the largest of the flames looked to be more than a hot dog roast away; therefore, I ventured my head up into the ragged and torn-metal escape opening. I saw fire trucks surrounding us as they started to spray foam and water on the cockpit. It was then that I realized that we were no longer attached to the wings and the rest of the fuselage had vaporized. I was amazed to see the scene that greeted my eyes. The engineless wings lay several feet behind the separated cockpit with huge flames leaping from the center wing tank. We still carried almost 75,000 pounds of fuel, so there was no shortage of combustible material. I waved desperately to the firemen as they deployed more hoses and equipment while others trained their nozzles on the flames. Rapid coordination between the tower and fire department had resulted in an incredible, life-saving response time.

Climbing back into the airplane, I threw an escape rope out of the newly blasted exit hatch above and in front of the EW's seat and returned to working with the RN to free the EW. Then, in spite of the Gunner and RN having broken legs, they both managed to climb out of the hatch and down the rear of the severed crew compartment.

We worked for several more minutes to safe the defense compartment ejection hatches and remove the EW and Navigator. The rescue crews handed me a big cutter which worked perfectly to sever the ejection seat arming tubes, thereby disabling the seat. The EW's injuries were severe - a badly broken right leg and serious leg and facial cuts. It seemed like the firemen were super-human as they lowered a sling and hoisted him straight up out of his seat, like there was a strong crane above us. Next, the navigator had to be lifted out in a similar fashion - both his ankles had been crushed in the initial impact.

With the main part of the fire at bay, the fire department was able to take its time using the EW and copilot's hatches to complete the rescue of the remaining three pilots. Careful work with back boards enabled firemen to safely extract them without further injury. All three had in fact sustained serious spinal injuries in the crash. As I sat down in the back of the ambulance with the RN on a stretcher next to me, he painfully said, "Welcome to K.I. Sawyer!" I returned a thanks and assured him that this welcome couldn't be topped, nor should it be.

The resulting 1000-page report containing causal findings from the accident investigation board revealed that the aft body tank was missing a flame arrester in one of several fuel pumps. Since the tank capacity is 45,000 pounds of fuel, it is one of the largest single tanks; however, it contained only 1,500 pounds by this time in the sortie. Thus significant room remained for fumes to build up and for some fuel pumps to be uncovered, exacerbating the potential of interior ignition. So, when a spark from the unprotected electric motor in the pump came in contact with fuel vapor, it ignited a chain of explosions that started by blowing off the aft section of the aircraft, removing the entire empennage, and causing the aircraft to immediately tip nose-down into the runway. The three biggest pieces remaining were the aft wheels attached to a small piece of fuselage, the wings which were mostly intact, and the cockpit, which ripped itself from the fuselage just forward of the front wheels.

The blessing is that all eight crewmembers survived. Despite getting in the predicament in the first place, God was watching out for us that night. Had the accident happened several seconds later, we would have been higher, faster, and out over the trees and forests of Michigan's U.P. Had it been on final approach, the results would have been equally deadly. Also, since we were landing to the south that night, the crash occurred as close as it possibly could have to the fire department. Touch-and-gos to the north would have put us almost a mile and a half away from the fire and rescue professionals. Everyone had broken bones, except for my questionable minor fracture, several cuts requiring stitches, and bruises. For once, my poor posture may have saved me from sustaining a broken back. It was very sore, but at least I could walk. The two pilots in the seat along with the instructor pilot sustained broken backs and some degree of permanent paralysis. The offense (navigator) team both had broken legs and/or ankles, while the defense team (EW and gunner) both had broken legs. I was mostly so covered with bruises that I felt like I'd just played my first high school football game - twice.

The other three pilots were permanently grounded and two regained the full ability to walk again. The offense and defense teams all were later cleared to fly, but only the navigator remained on flying status for any significant time, staying with the B-52 until he retired only a few years ago.

When I first wrote about the accident many years ago, it was for Strategic Air Command's Combat Crew magazine. I talked about several things that could be learned from this "violent and potentially fatal accident." I emphasized knowing your aircraft's safety equipment, its location, and how to use it as well as encouraging the reader to periodically go over the locations of each fire extinguisher on the airplane. You may remember the crash axe and its use for opening a stuck entry hatch, but I tried to remind you that it may be useful for several things in an accident. Another emphasis item was that each person should take some free moments occasionally to look over hatches, doors and windows and assure yourself that you could get them open even if it's dark and the airframe is slightly twisted. Of course, a new pet peeve of mine was strapping down the communications kit and other heavy or bulky equipment to the defense instructor's position or another secure place.

That night, as I worked to get out of the aircraft, nearly suffocating, it did not even cross my mind to grab a portable oxygen bottle or hook back up to the emergency bottle on my parachute. I also did not know the location of my flashlight. Obviously, it had been thrown about upon impact, but having my small Maglite handy in a pocket would have made a world of difference in such darkness. I was very fortunate that the crew compartment didn't roll after we departed from the rest of the aircraft. If we had rolled, my injuries without my seat belt buckled could have been much more serious.

Finally, this is a memory to tell with my children and their children if I get the chance, but it will always be a story for me of my life being saved, and other lives being spared. It encourages me to make the best of every day. My years in the B-52 built a rich storehouse of lasting memories. They are filled with fun and challenge as they continue to influence the shaping of my character as an Air Force officer. I wouldn't trade them for anything. It's amazing to me that my children and possibly their children might still be flying this best bomber ever built.

Resting in mud off the end of the runway.

The Last Flight of BUFF 574 -
"Home At Last"
Gary Miller

We were inbound to Griffiss Air Force Base on final approach. The evening sky over upstate New York was black and heavy with rain. The lumbering B-52G bomber, #59-2574, was about five miles from landing and inside, ready to get home, was the crew: Capt William C. Pool, Pilot; Capt William R. Fox, Copilot; Capt John H. Moran, Radar Navigator; Capt Robert G. Miller, Navigator; Capt William H. Bengle, Electronic Warfare Officer; TSgt Evaristo Vasquez Jr.,Gunner.

The training mission had been a higher-headquarters directed Operational Readiness Inspection – an ORI. Outside it was dark and raining, and inside we were all behind on the landing checklists, so the "pucker factor" was up - that is to say, the pressure was on. The mission had not been a smooth one; we had fought it all the way. Somewhere 500 feet above the ground over Kansas one of the engines had malfunctioned, becoming uncontrollable. The copilot, Captain Bill Fox, who had only recently checked out in the aircraft, had been instructed to shut down the engine. Losing one engine out of eight was

125

not a serious problem, but little did we realize that same engine would later come back to life with disastrous results. At the time we pressed on, still at the low altitude of 500 feet above the ground, heading toward our targets. It was 8 May 1972.

The crew had flown only once before as a crew – which was not the optimum way to go on an ORI mission. All the Griffiss crews were in a state of continual change as everyone adjusted to deployment cycles to Southeast Asia for Arc Light missions over Vietnam. As a new crew we all realized that our teamwork, essential to safely flying a big eight-engined bomber, was not as finely honed as it should be. But we were hacking it; we had gotten the job done - all bombs on target - and that would translate into no reprimands by the Wing Commander. Just a few more minutes and we would be home free.

As the aircraft descended, less than three miles out, the pilots had the airfield in sight and could see the runway lights through the windshield wipers and driving rain and fog. Downstairs in the pit, from the navigator's station, I watched every sweep of the radar and monitored the decreasing altitude. The yellow strobeing of the radar and the red lights given off from the instrument panels gave the cramped lower nav / bomb deck an eerie glow. As we descended from our low level altitude, I noted as always that we had passed the altitude where neither I nor the radar navigator who sat beside me could safely eject from the aircraft if anything went wrong. The downward ejection seat in which I was riding was a death trap at low altitudes. The probability of survival below 500 feet above the ground was not good. Trying to relax, I leaned back and watched the altitude decrease as the radar painted the base and the runway. We were all tired after more than nine hours in flight. At any rate in just a few minutes we would all be home.

Over the runway I could feel the aircraft flair and start to settle out for the landing. As the pilot cut back the throttles however, the roar of an engine remained on the left side of the plane. "No sweat," I thought. The new copilot must have forgotten to turn off the engine on the AGM-28 missile which hung down under the wing. Missile engines were often used as extra thrust during flight. The giant plane floated down the runway…floated…floated…floated until I sensed something was not right. Nope! "This is not good!"

Upstairs the pilot, Capt Walt Poole, was fighting with little success to get the bomber on the ground. The engine that had been shut down at low level had come back to life and was starting to run away at maximum power, taking the giant bomber further and further down the runway like a run-away freight train. Walt was unable to stop it. He tried to deploy the drag shoot in an attempt to slow down. It did not deploy! No shoot! Downstairs I knew things were not right as I waited and waited to feel the landing gear contact the earth and the brakes engage. I looked at the airspeed - 150 knots. If we didn't stop soon we would run out of runway. That was when I heard Walt say, "We're going around." Immediately I could hear the sound of all the engines coming to life and the giant bomber began to respond. Just as suddenly, the engines went silent - except for the run-away engine on the left. It continued to force us down the runway, the landing gear barely touching the ground. Walt, his voice firm but still in an even tone, announced, "Crew, we are going off the end of the runway."

"Great! Just great! I am in the belly of this airplane, the very bottom. What am I doing here? It's 11:00 p.m. I'm supposed to be at home...in bed with my wife, or out of the Air Force. If only I hadn't extended my tour of duty. Damn! How do I get out of this ejection seat? Maybe I can unstrap, climb upstairs. Maybe have a fighting chance. No; probably wouldn't make it up the ladder in time, be thrown forward. Bad idea."

I looked over at Capt John Moran, the radar navigator. He definitely understood the seriousness of the situation, and being Catholic, was crossing himself. I signaled to him to safe his ejection seat, lock his harness, put on his oxygen mask and pull down the helmet visor. I did the same and put on my flight gloves, throwing all my navigation gear - charts, manuals, and checklists - on the floor behind me. I pushed in the desk, leaned back and braced myself. I knew our chances were not good. I recalled that all the accident reports I had read on B-52s attempting crash landings had fatal results - too big, too much fuel, and too many engines.

But as I watched the radarscope go around, and then around again, calm settled over me and I waited for the inevitable. The green - yellow glow I saw through my shaded helmet visor gave me a surreal vision of the last moments inside the dying bomber.

The huge aircraft was racing toward the embankment at the end of the runway. Wet with rain and reflecting the red strobe lights on the wings, the runway was being used up quickly, and suddenly the aircraft went off the hard surface with a mighty lurch. With the sound of grinding metal, the landing gear sheered off the careening aircraft, and then the two large AGM-28 missiles that hung under the wings contacted the ground and were ripped off, cart wheeling behind. Appearing to fly as it departed the runway, the BUFF seemed to remain suspended for a split second, then fell out of the sky, hitting the muddy earth. The impact resulted in the aircraft breaking in half just behind the crew compartment and skidding down the embankment with the outboard run-away engine still whining.

Inside it felt like someone had pulled the rug out from under me as the aircraft fell and broke in two. The radarscope flashed bright, just like a light bulb sometimes does just before it burns out. Then everything turned as black as pitch and went deadly silent. I could not believe it. I was still alive. No explosion! In the dark the first thing that started to register was the now familiar sound of the run-away engine on the left side…then the smell of JP-4 aviation fuel. "Our luck may be running out, and fast," I thought.

As I hurriedly unstrapped and disconnected my oxygen, I heard movement upstairs. Capt Bill Bengle, the electronics warfare officer, and Sgt Evaristo Vesquez, the gunner, who were sitting upstairs aft of the navigations compartment, were trying to get the hatch open so we could escape the dead bomber. I waited and waited but could only hear sounds of straining, pushing and shoving. The smell of JP-4 was getting stronger and the engine was still running. "Hey, Bill! You guys want to get that thing off before we all become crispy critters!" The adrenaline must have kicked in as a rush of damp cool air came in and I heard the hatch open.

Climbing quickly up the ladder, I could see the pilots in the dim light standing in a crouched position on the flight deck. As I emerged through the open hatch onto the slick surface of wet aluminum, I could plainly see twisted metal in the large break behind the crew compartment which now separated the fuselage into two pieces. The gurgling of the aviation gas was very audible, as it leaked out of the wing - and the hot engine was still running. When I looked up, I could see the fire and rescue trucks at the end of the runway, red strobes flashing, unable to reach us from the top of the embankment. They might just as well have been miles away! No time to waste. Hand-over-hand down the escape rope from the top of the airplane, helmet still on, my feet touched the ground. Two of the crew were already on the ground. We took a rapid inventory of each other - no fatalities or injuries, but we were lucky! But at that moment, our only focus was to get as far away from this one as possible. We sprinted away from the old dead bomber as fast as our legs would carry us through the ankle deep mud.

We learned later that a Boeing representative stated that most likely the only reason the aircraft did not explode on impact was due to the mud caused by the recent heavy rains. The sight of the gigantic bomber split open, revealing twisted metal, and partially submerged in muck, made us all wonder if we would ever want to get back inside one of those things again. We all did eventually - some of us reluctantly, but as a crew we never flew together again. A few months later some of the crew faced death again in Linebacker II missions above Hanoi.

Looking back, despite all that happened, it was once again - a completed mission. Home at last.

Four Klaxons in 24 Hours
John Morykwas

Many people do not know the seriousness, or the stress placed upon Crewdogs in their chosen line of work. This refers not only to bomber and tanker Crewdogs but to missile crew members and everyone else assigned to the Strategic Air Command, (SAC.) For SAC personnel, one second the United States of America is at peace with the world and, in the next second, the USA could be in all out nuclear war. Such was the case with my life in SAC as a B-52 Crewdog.

Like others, I served the famous seven continuous days on nuclear alert (from Thursday to Thursday) with my crew, Crew E-12, of the 7th Bomb Wing, in the 20th Bomb Squadron at Carswell AFB, TX. On one such tour an occurrence took place that will forever be etched into my memory bank. Crew E-12 was still in its infancy, sharpening its crew's abilities when our copilot was not able to join the crew for our scheduled alert tour. Instead, a substitute newbie copilot was assigned to the crew.

Crew E-12 consisted of the Pilot, Captain Bob "Buffalo Bob" Sanford; RN, Miguel Primera; EWO, Lynn Wakefield; and as Navigator, your's truly, 1st Lieutenant John Morykwas. I do not recall our Gunner's name for that tour since we changed them so often. It just so happened that alert tour was held during a time of heighten tension within the world, but for what reason, I can't remember.

During that time, someone within SAC wanted to test, and/or show the abilities of SAC to respond to a national emergency, by having a few "practice" Klaxon alerts to sharpen up the Crewdogs. At approximately 0200 Carswell time, that is 2 AM in the morning, the Klaxon sounded and the crews responded as quickly as possible to their aircraft and called in their ready times.

During the morning intelligent brief, the Crewdogs were informed what was happening in the world, which probably resulted in the Klaxon that early in the morning. The Crewdogs were placed on a heighten level of alert. The official, unofficial orders restricted the crews to the alert facility, so this put everyone on edge.

At around 1000, that's 10 AM, Carswell time, the Klaxon sounded again. That time we had some of the 7th Bomb Wing, higher-ups, evaluating the response in person. I think the Crewdogs previous response time to the 2 AM exercise did not please SAC. The Crewdogs knew the morning test was an exercise, since the Wing Commander was seen loitering around the Alert Ramp.

One must realize that every Alert exercise involves a lot of work to re-cock the aircraft for alert status, doing such things as topping off the fuel. They are no picnics for either the Crewdogs or for maintenance personnel.

At about 2100, 9 PM. Carswell time, there was another Klaxon. That was our third Klaxon within 24 hours; SAC was really working the Crewdogs. Again the crews responded, and again found that it was an exercise. One must remember that every Klaxon was ultimately treated as if it is the real thing. It's like a game of Russian Roulette, you never know if the next time the trigger is pulled whether or not a real bullet will be in the chamber. For every Klaxon a Crewdog's heart winds up in his throat, since it was treated like it was the real thing, until told other wise. There is always a sprint to the aircraft, and a race to have the BUFF ready to do its thing. For some Crewdogs, it may be a half a mile sprint driven by an adrenalin rush. It's Pavlov's dog response, except that in this case, it's a SAC's Crewdog response. The response to the Klaxon is automatic, and each and every Crewdog has one thing on his mind, the mission. It is automatic. Crewdogs are trained steely-eyed-killers. It is their profession.

The talk throughout the alert facility was about the number of Klaxons held within an alert tour. None of the Crewdogs had ever had more than three Klaxons within one tour, and some tours had no Klaxons. We were thoroughly beaten that day. Very few of us had experienced even three Klaxons within a whole tour, much less three within 24 hours. The Crewdogs were thoroughly exercised that day, and tired. No one expected another Klaxon for the tour, much less for the remaining 24 hours.

At approximately 2400, midnight Carswell time, the Klaxon sounded again. I think everyone thought that was the real thing. As all of us responded to the Klaxon, the race was on to exit the alert facility to man out BUFFs. It just so happened that one of those famous Texas thunderstorms was passing through Carswell at the same time. In the morning briefing, the Alert crews were informed of the weather and the time it was expected to hit.

The alert facility crew quarters are located below ground. Since the alert facility was built like a bomb shelter, little sound penetrated the sleeping area. The crews are housed in the following order, Pilot and Copilot in one room, the Nav, RN, and sometimes the EWO in another room, and the Gunners from two crews housed in another. For some reason, the Pilots' rooms are located closest to the exits. The bottom area of the alert facility has an incline ramp to ground level with two double doors separated by around a six-foot area. A response to a Klaxon while the Crewdogs are sleeping is like having a maximum response of college students in a dorm to a real fire. It is a mad dash toward the exits.

It just so happened, that the Copilot was about 10 feet ahead of me and my other crew members running toward the door. As the Copilot hit the first door there was a bright blinding flash, and a tremendous deafening boom. It was a lightening strike, and it was very close, if not striking the alert facility itself. The new substitute Copilot immediately dropped to the floor, and curled up in a fetal position. When I, and the RN, Miguel Primera, or maybe the EWO, Lynn Wakefield, approached the Copilot, and asked what was his problem, the statement out of the Copilot's mouth was, "They done did it, they dropped The Big One."

Knowing it was not, "The Big One," but only a thunderstorm, I, and the other Crewdog grabbed the Copilot by his flight suit, still in the fetal position, and carried him out the door to the alert truck. We lowered the tailgate, and threw him in the back of the bed. When we entered the alert truck, the Pilot asked "What's up?" all we could say was "Floor it!" The storm was one of those gully washers. A few seconds spent in the rainstorm resulted in a total soaking. About half way to the aircraft the Copilot finally came to his senses, and started beating on the roof of the cab saying "Stop, and let me in." No one stops an alert truck in response to a Klaxon, not even for a stop sign, red light, or security police - the first stop is the aircraft. When we arrived at the aircraft the Copilot's training took over, and he responded as a Copilot should, but was soaked to the bone. Had that been a real launch, I think the Copilot would have frozen to death at altitude, but again, it was an exercise.

That was a record, four Klaxons within 24 hours, and a lesson learned by the Copilot to be alert during the weather briefing. The next day our regular Copilot reported for alert. I do not remember the substitute Copilot's name, but it's one of those experiences that you always remember what happened, but forget who the individual was. Maybe this was a condition to protect the career of the individual. I never saw the Copilot again at the alert facility or in the squadron building.

At the time, it was a good laugh, but in reality, it was serious. It was an example of the psychological pressure and stress placed upon a Crewdog. It makes one look back at the life of being on Alert, and remember how it felt to not know if the next Klaxon was not an exercise, but the real thing?

Chapter Three

Southeast **Asia** [south-eest] [ey-zhuh] – *noun* – The countries and land area of Brunei, Burma, Cambodia, Indonesia, Laos, Malaysia, the Philippines, Singapore, Thailand, and Vietnam.

Reflex Alert on the Rock – 1964
Nick Maier

The B-52 Crewdog stories began for me in the early 1960s at Castle AFB. I was one of the "homesteaders" stationed there for almost seven years, beginning in 1956. SAC finally tracked me down, among a multitude of others, and by 1962 I was a B-52 aircraft commander in the 330th Bomb Squadron. The Cuban Missile Crisis delayed our squadron's movement to March AFB in Riverside, California. The rumors of the B-52 replacing the B-47 on Reflex Alert at Andersen AFB, Guam, were also forgotten with the grind of flying back-to-back airborne alerts during the Cuban event. Reality set in very quickly, as we parked our B-52s into the alert lines at March in early 1963. The word was to pack our swim gear and zorries, we were heading for Guam.

I had mixed emotions about flying at 35,000 feet across the Pacific Pond, to pull Reflex Alert on a tropical island 6,000 miles from southern California. There were many scenarios that I encountered on reflex deployment flights to Andersen AFB. An hour after take-off, a 20-minute arm-building exercise was scheduled, topping off the bomber's fuel tanks during in-flight air refueling through a wall of thunderstorms just west of the Golden Gate Bridge. The instant that my sweat soaked flight suit became reasonably dry signaled that it was then time for me to accomplish one of the many critical enroute procedures, each prompted by ironclad mission timing. I celebrated the moment with an undercooked TV dinner at the navigation checkpoint 200 miles north of Hawaii. I munched melted, sticky candy bars directly over Midway Island while marveling at the sight of the Pearl and Hermes reefs drifting past the bottom edge of my window. Finally a battle with diarrhea prevented me for the third time in a row from seeing Wake Island. The memorable part of the journey was the 20 tasteless menthol cigarettes I inhaled, and the jerry can of GI coffee I drank to stay awake.

Within hours after attacking Pearl Harbor, the Japanese sent their bombers stationed on Saipan to begin a two-day bombardment of Guam. By December 10, 1941, the outnumbered American Marines and Guam Insular Forces were forced to surrender. The island became part of the Japanese "Greater East Asia Co-Prosperity Sphere." The new occupiers of Guam changed the island's name to "Omiyajima," and immediately began to build up the points of defense the Americans had already surveyed. Using entire families of Guamanians as slave labor, the Japanese chose Orote and Dededo as their primary sites for airfields. When the Americans finally returned in 1944, these same airdromes became prime targets for both the Navy carrier planes and the P-47s launched out of the recently captured island of Saipan. By June 20th, enemy air power was finished. The destruction of airfields on Guam was a direct contribution to the successful "Marianas Turkey Shoot," in which the entire remaining Japanese Pacific naval air arm was totally destroyed. The home islands of Japan were now within striking distance of Curt LeMay's 20th Air Force.

On August 10, 1944 the American invasion forces made a formal announcement that organized enemy resistance in Guam had ceased. It would still be weeks before the last of the sake-induced banzai attacks signaled the end of the war for over 50,000 Japanese military. Even before the sniper bullets stopped ricocheting off their armored bulldozers, the gallant Navy Seabees cleared the rubble of burned and twisted Japanese Zeros and Bettys to make way for the advancing Army Air Corps. In a matter of a few weeks, three airfields were constructed on the fairly level plateaus located on the northeast corner of Guam; Harmon Field, a depot and maintenance base; Northwest Field, a fighter base; and North Field, a facility for the mightiest of the new American bombers, the B-29.

The advent of the air age opened the floodgates for endless streams of Gis and their flying machines. Harmon and Northwest fields were closed shortly after V-J Day, and all air operations were consolidated above the cliffs of Pati Point at North Field. The canvas tents of the re-occupation gradually gave way to more permanent corrugated sheet metal Quonset huts, which became the native's principle source of housing material, immediately after any typhoon passed through. Following an already established Air Force tradition of naming air bases after the heroic death of either a local born airman or an aviator who had become notable for any other reasons, North Field was renamed Andersen Air Force Base. This act was cast in bronze

April 15, 1950, in honor of Brigadier General James Roy Andersen, reported missing on a flight from Guam to Hawaii in early 1945.

Before the last B-29 aircraft left "the Rock," SAC had already deployed their B-36 Peacemakers. In 1954, an entire wing of B-36s from Fairchild AFB, Spokane, Washington, arrived and initiated a rotation of weapon loaded aircraft, on alert and ready to launch. This birth of the Reflex alert concept began an unending chain of stateside aircraft and aircrew deployments and re-deployments. Within a year, SAC's Third Air Division assumed command jurisdiction of Andersen from the Far East brown-shoe Air Force. By the end of 1959, SAC had retired its last B-36. The replacement B-47 meant that the Command was now the first pure jet combat bomber force in history. The flight time to Andersen was jetted in half. Sheet metal Quonsets were discarded for concrete, typhoon proof barracks, and the construction of dependents quarters was an unmistakable sign of permanency. Andersen was well on the way to becoming civilized, in spite of the miseries of weather, winds, and mildew.

When SAC announced retirement of the B-47 fleet to an Arizona desert bone yard, there were serious concerns to this prospect among the worrisome populace living on this hub of Micronesia. Their non-voting representative to Congress had hardly gained the floor for his impassioned speech, extolling the strategic importance of his island and the certain economic doom of his constituents, when rumors became a reality. Since the B-52 was rapidly replacing the aging B-47s role in the Cold War stalemate of the 60s, it was only logical that the eight-engine behemoth would also fill the empty alert parking stubs at Andersen AFB. The first aircraft to arrive were also the oldest models in SAC's inventory. B-52Bs flew non-stop from March AFB in southern California and Biggs AFB in west Texas. They were immediately topped off with fuel, uploaded with thermonuclear weapons, and parked wingtip to wingtip on the former B-47 alert pad. Their blunt noses were pointed to the northwest, toward the only targets of strategic, political significance on that side of the globe.

To this day, I sometimes open some hung-up doors in my sub-conscious, which   produces a full color, wide screen montage of flashbacks to life on Reflex alert at Andersen. There is the long, straight alert line of slightly tarnished, aluminum hued B-52s, with their pure white bellies, forever out of focus, shimmering in the waves of equatorial heat rising off the concrete alert pad. An ugly string of concertina wire, with a surly GI guard at the driveway entry point, converted a three story cement barracks and its nearby mess hall into another SAC alert "hole." Any similarity to the dozens of SAC alert forces scattered elsewhere around the globe ended immediately after entry.

The windows of each crew room were covered with typhoon proof steel louvers, with a weeping, foul smelling air conditioner extending out partially blocking the walkways. The total humming sound of the coolers resembled a massive beehive, and the constant overflow of rusty water permanently stained the concrete terraces. Outside on the grassy yard under the tropical sun, a country club atmosphere prevailed, with the half naked crew members burning their bodies in stages, from rare to well done, while either playing games, playing the radio, or playing dead. The gunners were the only troops with enough sense to stay out of the ultraviolet heat, congregating instead in an air conditioned room to play 'round the clock pinochle, honing their skills to hose the stateside tournaments.

The party always ended when the totally unfamiliar sound of a Navy Klaxon horn signaled an alert. Guam's U.S. Navy landlord had decided to install shipboard Klaxons at Andersen, rather than send for standard stateside Ma Bell alert horns. In spite of the fact that the crewmembers lived in an atmosphere of constant anticipation of an alert call, the point of maximum pucker always came with that first blast of sound. I was always amazed at the slapstick drill of responding to practice alerts at Andersen. I also remember that since swim suits and sunglasses were the standard alert uniforms of the day inside the crew compound, I had never answered the call of the Klaxon wearing anything else. Each exercise produced a mass exodus of sunburned cursing men, running toward their vehicles lined up for a Le Mans start, everyone yelling and screaming, "Dive! Dive! Dive!"

I often wondered if it was more than a coincidence that every Coco exercise, which directed taxiing the entire alert fleet down the runway simulating an EWO launch, resulted in an automatic response by Mother Nature to purify the operation with torrents of blowing, tropical rain. By the time the engines were cartridge started and the airplanes were released to taxi by the thoroughly drenched ground crew, the wind and rain had reduced visibility to a mere 100 yards. Each great, noisy machine with its tall silver shark's tail would burst out of its lair, blowing ramp water back vertically into the air off the blast fences, which fell once again on the soaked crew chiefs, who were scurrying about like drowning rats. Our bomber's 60-inch tires produced speedboat wakes down the taxiways, which had been transformed into canals, with each wheel on the verge of hydroplaning completely out of control. The pilots were forced to lean forward in their seats, most of them still wearing their swim trunks, straining to see through the furiously moving windshield wipers which were unable to keep pace with the cloudburst rainfall.

To add surrealism to this farcical episode, the topography of the Andersen taxiway and runway system resembled a section of Route 66 through New Mexico, with hills and valleys that were unbelievable. To complete the frightening scenario, it was almost impossible to see the aircraft ahead, except for a moving blob of darker gray space. I was constantly called upon to rotate my handful of eight throttles through the complete 120 degree arc of the throttle quadrant in order to control speed, maintain spacing, and protect the lives of everyone within a couple of thousand yards, in case of a catastrophic collision.

Invariably, moving down the roller coaster runway and returning to the parking stubs after termination of the exercise would also bring out the equatorial sun. Within seconds, my crew would demand in unison that the pilots open their sliding windows for relief from the horrendous sauna conditions in the forward compartment. To add to our discomfort, each man knew our careers could end if the local one star division commander caught anyone deplaning after an exercise, wearing anything but a properly zipped up flight suit and fully laced combat boots. It happened only once; my crew completed a taxi alert exercise, and scrambled out of the unbelievably hot and humid aircraft with our bodies golden brown and glistening with sweat. We were met by the current division commander, who promptly chewed out our collective butts, accusing us of beach party behavior and lack of proper motivation for the seriousness of our assignment.

After the blustering general recovered from his apoplexy, he issued the famous flight suit edict. Woe to those in the future who deplaned without flight suits and flying boots. A suitable punishment would be ceremonial sacrifice, being thrown off of Pati Point in memory of the legendary suicidal native lovers, who leaped to be dashed on the rocks 600 feet below. A far worse calamity was a promise that the unfortunates would be restricted to their quarters in full alert costume. The gunners never complained about not having windows in their compartments that could be opened for life renewing fresh air. They calmly sat in the pools of their own salty sweat, and whispered in a disparaging sing-song over the interphone, "Officers-are-candy-wimps ----" I still smile at the memory of the look on the general's face. I also have a corresponding involuntary shudder, recalling the endless hours of post alert ramp time I had logged with the crew, waiting for maintenance to patch up our broken bomber. During those periods of utter boredom, the crew's entire motivation was to re-cock our alert bird and get back to the "jungle rules" volleyball game that had been interrupted by some nameless SAC bird colonel in the middle of a frigid Omaha winter night.

There was one compensation for enduring a seven-day alert tour on the Rock - the prospect of getting a seven-day Rest and Ruination trip to Tokyo on the Yakota Flyer, along with the C-118 full of GIs and their families. We showed up immediately after getting off alert, usually still in our green bag flight suits. I believe this was the first time we all became known as "Crewdogs!" The R&R stories will have to wait until the next WWCD volume. The success of the Tokyo shopping trips was measured by the amount of "goodies" that were loaded into the BUFF bomb bay cargo box for the trip home to "the world." Those trips didn't happen until another seven-day alert cycle which was a blessing to rest and detox from much too much sake, and an occasional Hirohito's revenge.

By early 1965, Andersen Air Force Base had become a quiet, sleepy nuclear outpost in the far western Pacific. The country club atmosphere prevailed for the alert force aircrews and their supporting 3rd Air Division staff. The Guamanian attitude of mañana had also affected the headquarters brass, who had become tropical haole expatriates, and whose sole concern was the occasional typhoon. Hafa Adai, and "Where America's day begins," were the only local expressions that required memory space. A 40-ounce jug of Beefeater Gin was less than two dollars and a good year of Chateau Nuef de' Paup French wine was only 90 cents. Every beach on the island was a travel agent's dream, and the ocean temperature never varied. Life at Andy had its compensations. Compared to the plastic life in California, it was stepping back into a tropical neverland. The airbase sprawled over the northern tip of Guam, and other than a few vintage 1950 buildings, it hadn't changed much since the post war days. There was still a weather-beaten hand painted sign over the ramshackle Base Ops hut, proudly announcing to any newly arriving airman "Guam is

Good." I had often wanted to fly to Saipan, just to see if there was any truth in the rumor that the sign over their terminal said "Saipan is Better!"

The musty antique Andersen Officer's Club still echoed the ghosts of B-29 crews. There were several recent indications of a gradual buildup of base facilities. Most notably was the construction of a new MAC passenger terminal, and a control tower that was destined to become the highest such structure in USAF. I had been watching the endless building projects for an entire year. No one seemed to be in any particular rush to get anything completed, after all there was always tomorrow. It was a year later to the day that I had brought one of the first B-52 crews to that far-flung Pacific alert pad. I had an enviable assignment, stationed at a California SAC base considered to be "fat city" among combat crewmembers. In spite of the many personal pleasures I enjoyed in Hollywood, Hawaii, and the Far East, I had always questioned the rationale of flying 6000 miles to that god-forsaken place to pull hard alert. Every flight across the pond had acquired the monotony of a commute, where a minor emergency was a welcome distraction. The island's odor of decay and mildew was on everything I owned.

The day that trouble came to Paradise is still very vivid to me. I was sacrificing my body lying in the sun on beautiful Tarague Beach when the background chattering became eerily quiet. Everyone on the beach had stopped and stared off into space, pointing silently to the northwest horizon over the towering massif of Ritidian Point. A B-52 had completed its let down approach to the island, and was noisily entering into the downwind leg for landing at Andersen. The event was of no specific note on any other similar day. What was different, and to be forever painfully etched on every sunburned brain, was that particular B-52 was unlike any others that had previously journeyed to Guam. The gleaming white belly of the BUFF, which was intended to deflect the tremendous heat and radiation from the megatons of thermonuclear weapons that it was to deliver eventually, was completely sprayed over with an ugly black coating. I had heard that they were coming. On our next rotation stateside, my crew was scheduled for training in "conventional weapons delivery." It was obvious that SACs mission of nuclear deterrence had definitely taken a back seat to the mobilization of the Arc Light B-52 conventional bombing forces. It seemed only a matter of a few days, when we and our fellow crews were scheduled for re-deployment. As soon as the final Reflex alert lines came off the pad, I noticed an increase in the urgency to get the nuclear weapons downloaded and the aircraft deployed back to the states.

As I taxied my ancient B-model bomber through the slalom of Andersen's rolling taxiways, I passed sights reminiscent of WW II days. Huge earth-moving equipment had been painted a peacetime yellow, devoid of armor plate. Native Guamanians, instead of GI Seabees, drove them wildly in the steaming tropical heat, hacking out spaces at the edges of palm tree jungles to park the constant flow of weirdly painted B-52s. Stripped-to-the-waist sweating GIs were trying to make some sense out of the erector set steel revetments that had miraculously arrived by ship the day before. By the time their superhuman task was completed, innumerable three-sided steel bunkers would dot the huge expanded Andersen ramp area. Everyone secretly prayed that they were designed well enough to prevent a series of sympathetic explosions with any surrounding bombers, should a butterfingers weapon loading team literally drop the ball, and trigger their 225 ton firecracker. The arriving B-52s were parked into newly constructed steel revetments as fast as they were completed. The aircraft presented a crazy quilt of paint jobs. Some were completely black bottomed, others with only the underside of the wings painted black, and one completely black, looking sinister.

Returning our creaking, vintage B-models to California meant more than changing laundry for some of my squadron mates. A dozen of them were given invitations to tea with the bomb wing CO, during which they were informed that they had been tapped as warbird drivers in the Southeast Asian combat air action. Their participation in the Cold War had terminated. They suddenly found themselves exchanging their sky blue ascots and neatly pressed SAC alert force flight suits, for the wrinkled, over washed green bags and camouflaged helmets of crew members flying some other air museum candidates; the F-105 Thud, F-100 Super Sabre, B-57 Canberra, EB-66 Destroyer, or H-43 Huey helicopter. The luck of the draw during that infamous SAC tea party sent some of the winners to partake in the oldest craftsmanship of the air warfare age, the forward air controller. These conscripted FACs would suddenly find themselves zooming about the jungle treetops and down among the picturesque rice paddies. They were strapped into some of the oldest propeller driven aircraft still active in the USAF inventory, the A-1E Skywarrior and the O-1 Bird Dog.

The toll of losses among my squadron's hand picked air combatants began almost immediately. The casualties were reported by rumor, letters from surviving families, or death notices in the Air Force Times. A phone call in the night from my former reflex alert copilot's wife broke the news that he had been shot down near Vinh, North Vietnam in his F-105. He was listed as an MIA, and his body was never recovered. His tragedy was compounded soon after with the birth of twin daughters. As the F-105 strikes against North Vietnam intensified, the tally of KIAs, MIAs, and POWs read like a past roster of my bomb squadron. The Time's obituaries included former B-52 aircraft commanders, copilots, Ews, and an occasional navigator, lost in some combat action over obscure villages and seaports that became household names in funerals across America.

This story is dedicated to all these former Crewdog warriors, with a special remembrance for my copilot Captain David J. Earll.

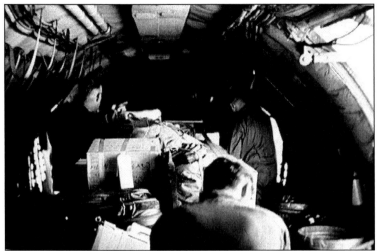

Riding as a PAX on a 135.

Early Arc Light
Peter Seberger

My first assignment out of pilot training was to the 20th Bomb Squadron, then based at Barksdale AFB and flying 15 B-52 F models. The other units flying F models at that time were based at Carswell, Mather, and Columbus as well as Castle, which used them for initial training in all models for crew assigned to the B-52 force. I had been sent from Pilot training at Vance AFB, Enid, Oklahoma, directly to Barksdale where they cut new orders to send me for initial Combat Crew training. I flew the first four missions at Castle in B models and the last seven in F models, and was "initially qualified" when I left in April of 1964. I then took survival training at Stead AFB near Reno, followed by a nuclear weapons school at McConnell AFB, Kansas. When I got back to Barksdale I was locally trained in the GAM-77 (later designated the AGM-28) Hound Dog missile, given a couple of training flights, a local flight checkout, and then several days intensive study in the war plans and positive control areas. It was July before I was finally assigned to a combat ready crew and ready to carry my weight.

Since we primarily carried the B-53 and the missiles carried the B-28, I did special weapons training for those initially. The wing did not begin conventional weapons training until that fall, when the Vietnam War plans involving B-52 participation filtered down to the local units. At that time in SAC, most of the Aircraft Commanders and many of the crew were fairly senior Captains, Majors, and not a few Lt Colonels. Most of them had been pilots during and immediately after World War II and Korea, so they had already had extensive conventional weapons training previously in their career. About that time some of the newer conventional weapons, like the leaflet and the cluster munitions dispenser were introduced. I remember the weapons officers who conducted that training multiplying the number of BBs in each bomblet (155, I think) by the number of bomblets carried in each vertical shaft of the dispenser by the number of shafts and coming up with some millions of BBs each plane would carry. They also, along with most of the crew force, laughed at the idea that the capabilities of the BUFF would ever be wasted with only conventional munitions. Little did we know?

Later in the fall of 1964 it became evident that the planners were serious and Barksdale was tasked to practice the enroute and conventional weapons delivery tactics then envisioned. I have seen reports from crew members who were at Mather during that time who claimed they did the initial testing and development of those procedures, and they might well have. I was a junior officer and neither curious nor in position to know about other unit involvement. My crew flew to practice enroute cell, a little fingertip, and low level cell with what we considered pretty minimum separation. Conventional bomb delivery tactics envisioned at that time included 6-800 feet AGL and two miles between planes at low level, with a "short look" (a tactic required for delivery of certain nuclear weapons) maneuver to 2,500 feet AGL for bomb release. We were to carry 27 internal and 24 external 750 lb. class munitions on each airplane. Planners envisioned low level because at that time the plans were to bomb the Hanoi and Haiphong areas at night and there was no air defense (we thought) to speak of, as the SAM sites were not yet being built.

We did have one crew that dropped a full load of 51 of the 750 lb. bombs at Matagorda Island bombing range to test out the theory. They came home with quite a few holes in the bottom of the airplane and wings, so that got the attention of the planners. I don't know when the

We Were Crewdogs IV

final decision to abandon low level was made, but it was well into spring of 1965.

In my military career I have seldom had a situation which was convenient for my personal life, but one such surfaced when I got to Barksdale. It so happened that Second Air Force headquarters was then at Barksdale along with the Second Bomb Wing, which had the 20th bomb squadron and an air refueling squadron as well. That was well before the availability of high speed secure communications, so there was a requirement for regular courier service between 2AF and SAC. A KC-97 from Barksdale made the round trip from Barksdale to Offutt on Monday and Friday mornings, returning in the afternoons. Offutt sent a similar airplane on Tuesday and Thursday mornings to Barksdale, returning in the afternoons. Barksdale bomber crews pulled seven-day alert tours and changed either on Wednesday or Thursday mornings. My crew was generally on a sortie which changed on Thursday, as Wednesday crews got an additional day off due to the fact that Sunday followed their CCRR. The senior and stan-eval crews liked that schedule, and I was never on such a crew. However, the Thursday crews bitched enough that the staff generally scheduled the following Monday off for the Thursday crews to even out the time off. At the time most bomb squadrons were staffed at 1.8 crews per airplane, so it was normal to have two weeks between alert tours and sufficient time for flying was available during that time. Later, as the war built up the strain on the crew force became evident and this ratio of crews to airplanes eroded back to 1.5, then even lower, and by 1966-68 back to back tours (seven days off between seven day tours) were common. Anyway, Thursday alert changeover worked fine for me as then I could catch the Offutt courier back to Nebraska that same afternoon.

I was a young lieutenant, not yet married and like many of us had decided that I could afford a new car. The first Ford Mustang had come out in 1964 and I was tempted to buy one of those but Chrysler also produced the Plymouth Barracuda, a Valiant derivative, about that time and I thought that car looked better. I bought one of those, (for $3,300) and then drove my old English Ford car to Offutt, and kept it parked in a lot near the passenger terminal where the KC-97 courier planes loaded and unloaded. My wife (then my girlfriend) was in Lincoln going to school so I could get a weekend in Nebraska, staying with relatives in Lincoln, and then Monday come back to Offutt to catch the courier back to Barksdale. If I had to be back by Monday morning, which was seldom, I could buy a space available ticket from Eppley Airport at Omaha for 30 dollars and fly to Shreveport on Sunday afternoon. On the few occasions I had to buy a ticket, I never failed to get where I needed to be. That situation worked great during summer and fall of 1964, but when we finally deployed that spring it turned into a minor headache, since the Ford had to be left at Offutt for nearly four months in the spring of 65. Believe it or not, it started when I finally got back to it, after sitting for nearly four months undisturbed. That was the first year of Chrysler's new five-year, 50,000-miles warranty, which required the oil to be changed every three months or 3,000 miles to keep the warranty current. During that spring of 1965, I had to have somebody who stayed at Barksdale take the Barracuda for an oil change after only seven miles since the prior change. That must be some kind of record.

An interesting sidelight, for those who missed it, was the privilege then in effect for military servicemen provided by all the airlines. They would sell a space available ticket which he could use for any open seat available just prior to departure. He had to be traveling at his own expense, be on leave or pass in a duty uniform, pay half coach price for the ticket, wait by the gate and have only carry on baggage which could include a hang-up garment bag and a small suitcase. Priority was by time of arrival at the gate, not sale of the ticket, so if a person missed one flight he would usually have priority for the next. I never missed a flight I wanted to be on, though I did get the last seat once. The airline would generally fill up first class last, but often those were the seats I got. My suitcase, left in a small closet by the entry door, was the three zippered canvas kit bag which I had been issued at pilot training, designed to be carried in the T-33 travel pod. It was/is (I still have it, though the zippers have finally failed and it is pretty ratty by now) about the size of the modern carryon now permitted, perhaps a couple of inches longer but no heavier. I was on the base pistol team wherever I was stationed during that period, and was early on told by the range NCO that (at least in Southern states) no peace officer ever would have a quarrel with a serviceman having a weapon while in uniform, so many of us did. I never carried it concealed, but in that kit bag under the outside zipper whenever I flew anywhere was my Model 27 Smith and Wesson .357 revolver, fully loaded with six hollow points. It was also the sidearm I carried during my Arc Light missions in 1965, though at that time I loaded it with solid bullets. As time went on, peaceniks and crazies began looting armories, and hijackings became so common it was no longer advisable to try to carry the weapon on planes. By that time I didn't travel much outside official duty anyway.

That summer and fall of 1964 was a pretty good time for me, as we were flying Chrome Dome, considerable flying, but also lots of alert and I didn't mind that since I used the time off pretty much as I described above. Early in 1965 though, tensions got higher as Rolling Thunder escalated in Vietnam and we gradually received more and more briefings and warnings about potential deployments. The weapons guys still did not think we would ever go. One of them said there were enough leftover WW2 bombs on Guam to completely equip 45 30-airplane B-52 raids, and they thought we would never use them up. Of course we know now that by 1966 they were almost out of those bombs and we wound up buying some back we had given to certain European countries to use the ammonium nitrate for fertilizer. Want to bet we paid a good price? In any event, early in February of '65 we were notified that deployment to Guam was imminent, and the crew force, and many of the maintenance and wing staff, were restricted to alert or one-hour home or telephone alert. The plan was that alert aircraft that were to deploy would have weapons removed, MER racks (empty) installed, and fueled to max and the crews would deploy with the plane to Guam. Planes not on alert would deploy likewise, with crews assigned as available. Staff, maintenance, and extra crew and complete crews who just did not have a plane to fly would ride in the tankers, who likewise deployed. I recall we only spent about three days on house alert before the deployment order came, on Valentine's Day of 1964 I think. My crew was not selected to fly a bomber, and flew over in the back of a tanker. It took about nine hours to get to Hickam for refueling, then another seven hours on to Guam.

When we arrived in Guam it was a real circus. The Bomb Squadron from Mather had deployed at about the same time as we did, and the arrival at Guam of 30 F-models, 30 tankers, and all the associated maintenance personnel made housing a premium, and it was a goat rope for a while. My crew was billeted at that time in the barracks building just across from a barracks called "Menopause Manor" because of the DOD school teachers who populated it. Somebody else could and probably has written a whole book about that situation, but I am not the author and this is not the place. At the time I was engaged to be married, and we had set the date a year earlier, to be June 19. It was the middle of February, so I wasn't worried as surely

149

we would be back - we were only TDY for goodness sake! I can't believe I really expected it at that time.

We had deployed in anticipation of action so immediately began preparations, briefings, and otherwise got ready for our intended mission. At the time I believe the plan was still to do low level night attack, so routes, tactics, and procedures got ironed out to do that. However, we were still not officially at war, and there were still several B-models pulling nuclear alert (similar to the old B-47 Reflex) on the base in TDY status. I know March AFB was one of the bases with B-models, there may have been others. Anyway we were not immediately executed so somebody got the bright idea that the F-models should also pull nuclear alert. Accordingly, they came up with a plan to have the alert F-models loaded with four B-28 special weapons occupying the forward and middle bomb bay stations, and nine 1000 pound class munitions on conventional hangars in the aft bomb bay. The Multiple Ejector Racks (MER) were loaded with 12-each M-117 750 pound bombs with only four MER electrical safety pins installed. The nine conventionals in the bomb bay had safing pins installed in the racks. In event of launch from alert the four MER electrical safing pins could be removed by the navigators during response, but there would not be time to remove the internal pins so the aft bombs were ballast. After takeoff the externals would be released, safe (at that time) over open ocean, and the mission would go non-refueled on nuclear targets particular to the sortie. When/if a frag-order for conventional strike was received, it was supposed to be quicker to reconfigure for conventional bombing, because of the time required and crew rest restrictions on the limited number of weapons loading MMS crews. They would only have to download the clip of B-28s and load two clips of conventional 750 or thousand pounders. At least that was the plan.

Crews on alert were given a vehicle and quarters near the flight line, and so it went. I remember a couple of Klaxon exercises while pulling alert, but I don't think we taxied on any of them. The flight line was really congested and Andersen was also the landing and refueling base for civilian contract and most Air Force military traffic which, in 1965, was really getting under way. An exercise which required us to taxi could have closed the runway for hours, and those aircraft transiting the Pacific would not have had the fuel or inclination to wait. There was a second parallel runway at Andersen, shorter and it was usually available for landing, but was too short for a max weight takeoff. I do not recall ever using it. Three times during that period between the end of February and early June the order to prepare for launch was given, the planes reconfigured, crews assigned, and we even loaded fresh water and what passed for MREs at the time on the planes. During our briefings on Guam I think we were told that JCS had recommended execution, only to have President Johnson refuse. Later (years later) I heard he was briefed to expect 30% losses and he balked, but I was never able to verify that.

The plan was still night low level delivery. That spring of '65 the SAM sites in North Vietnam were under construction and I think they were about half the targets - the power plants and port facilities were the rest, but I really can't remember that well. Suffice to say that as the SAM sites became operational and the murderous ground fire from small arms and AAA took its toll on the more agile tactical fighter-bombers who were getting hammered, cooler heads must have decided to abandon the low level idea. I can't say when that happened, but tactics were changed sometime around the time I went back to Barksdale to make a unit move to Carswell. In late May/early June the 9th Bomb Squadron from Carswell replaced the Barksdale unit on Guam and the 9th and the Bomb Squadron from Mather were tasked for the first sortie, on 18 June, 1965.

When my crew returned to Barksdale in late May or early June we had a few days to get ready for our move, so I drove my Barracuda up to Nebraska and left it with my fiancé so she would have transportation to prepare for our pending wedding. Then I retrieved and drove my English Ford from Offutt back to Barksdale. After I made the move to Ft. Worth into a rented house, my crew was immediately scheduled to pull alert through the date for my wedding which my fiancé and I had set a year and a half earlier. After considerable bitching to the wing staff and anybody else who would listen (and some that didn't), I was given enough time off alert to fly up to Omaha, catch a train out to the little town in Nebraska where I was to be married, and immediately drive back to Carswell, where I had to go right back on alert to replace the guy who replaced me earlier. I was then supposed to get two weeks of leave. YEAH, RIGHT!

While I was at Omaha waiting for the train I heard that the first Arc Light raid had been flown, with less than spectacular results. The losses on the first mission caused enough consternation that the whole plan changed. Since Carswell had 30 planes assigned, and many of the staff was already set up on Guam, it was decided to have the 20th Bomb Squadron, which had just made the unit move, return to relieve the Mather unit which had lost two planes on the first sortie. Some of the blame for the first mission fiasco was laid to inexperienced crews so the command was for the most experienced crews in the 20th to return first, and my crew was considered one of those. So while pulling alert before my scheduled leave, I was notified that I would get only three days CCRR, then have to go back to Guam. Of course, we were still not at war, so somebody else got my leave. None of my crew was happy, of course, but I was the newlywed and got to brief my bride on the changes. That was high on the "not-fun" list. The plan, I was told when released for CCRR, was for us to be gone for a week, and then return as a sort of reflex schedule was envisioned. A day or two before departure, I was called at home on CCRR and told that the one week was changed to three weeks, still promising leave on return. That briefing to my new wife was even higher on the not-fun list. My poor navigator did not get the word about the change to three weeks, so he showed for deployment with only flight gear, and he was pretty unhappy too, although I think his wife managed to get him some extra clothes before departure.

Shortly after arrival at Guam we were notified of a Frag order for another (the second Arc Light) strike, to commence on 4 July, '65. My crew flew that one, and another on the 7th of July, and after a week or so break from flying during which there were no new Frag orders, we were selected for the fourth and fifth strikes during the third week of July. All of those were 30-plane raids, takeoff around 2100 local, and 12-plus-hours duration. I remember the fifth mission (our fourth) was one on which our BNS Magnetron failed, and we switched leads to bomb on the right wing of the new lead crew. They failed to acquire the proper offset, so did not release, so we could not release either. Neither did the third plane, so we had a three ship coming home with all the iron. Not good! That was well before secondary target assignments were routinely made so we had to carry the bombs back to Guam. That required poststrike refueling because of the extra weight and drag of the externals, so Guam launched a strip tanker for our cell. That was the first and only time I ever had to use manual boom latching procedures, as our system failed to latch normally.

After leaving altitude for refueling we were committed to get some fuel or otherwise we would not make it back to Guam, even if we jettisoned bombs. After landing with the full load of bombs we found that the tanker in which we were scheduled to return to Carswell had to leave early because of a typhoon alert. We had to stay around anyway for three more days to debrief the brass about the reasons we had not released. That debriefing lasted two minutes since maintenance bought the problem. The lead plane over the target got a good grilling, though. During the three and a half weeks we were there, the overall plan changed yet again and the Pentagon decided that we should all just stay in Guam till December and rotate back as an entire unit. My crew and four other full crews had left Carswell with the understanding we would be back in three weeks. Many of them had house purchase papers to sign and moving to complete, so the A/C's of those five crews were able to convince the brass to let us return for 10 days to finish our move. Of course the penalty was that when we came back in early August we would be the last to leave in December. We finally left Guam after 24 days, and got 10 days minus travel time to settle our affairs at Carswell. The briefing to my new wife detailing on yet another change, no two-week leave, and the prospect of the next five months TDY was right at the top of the not-fun list.

Meanwhile, the Carswell finance office didn't do their job so my pay was "All F* U* but good", and my new bride was not happy. I had told her what and when I would be paid, and she had depended upon that. When my wife ran out of money she had to call my folks to wire some to tide us over, and that was really embarrassing to us both. Of course we got back on a weekend, so I couldn't do anything about it right then. My bride and I decided, spur of the moment, to take what would have to pass for a honeymoon some five weeks late. We went to New Orleans for a couple days, saw the sights, and tried to make the best of it as we had other plans that would just have to wait. "Welcome to SAC!"

It wasn't exactly SAC's fault; it was the accelerating war and Washington's insistence on trying to run it from there with only land lines and peacetime rules. SAC, of course, was reluctant to release any control over our planes and crews so there was some serious infighting among the major commands early on which I understand is the reason for interminable TDY orders instead of the one-year tours which became the Vietnam War standard. Anyway, the time came to redeploy and my mood was pretty sour. I remember showing up to out process and being shown a pay slip which I was supposed to sign to indicate that things were correct. Of course it was still AFU, so I refused to sign. That is when I gained a new respect for super sergeants. The airman I told that my pay slip was not right didn't know what to do, so he called a CMSgt over to try to work it out. He asked me what was wrong, and I unloaded on the poor man. I had four different pay adjustments which were supposed to have been made during the past three months, and I had visited the finance office to make those changes (for the second time) just that week. None of them were reflected on the pay slip. He listened to me bitch, looked at the pay slip, picked up a phone and started growling into it. After about 20 seconds he hung up, smiled, and asked me to wait. A few minutes later, a scared looking airman ran into the building with another pay slip, and the CMSgt asked me to sign it if it was correct. It was, I did, and my pay was always correct and on time from that day in 1965 until the day I retired in 1983, and ever since. I wish I had gotten that CMSgt's name.

So then we got on the tanker and headed back to Guam for what would be five months. I had rented a house in Ft. Worth from a retired Lt Col who had moved himself and his wife into an apartment in Dallas so she could run her beauty shop. I had spent fewer than ten days in the house when I left, and my bride had unloaded the U-Haul in which we brought back our wedding gifts and turned it in. While I was gone she was scheduled to begin a dietetic internship at Baylor Medical Center in Dallas. Part of the internship was the provision for a room in the student nurses' quarters, so while I was gone she seldom drove back to Ft. Worth. She was consequently somewhat out of the loop for squadron news. We depended upon letters for communication, since phone calls cost $4 per minute, which I found out when some problems arose with the house. But that's another story, complete with names and much cursing and invective, not suitable for polite company. Suffice it to say that I hope a certain pair of officers, one retired, from that incident will roast for a time in Hell for their sorry performance.

When we landed back on Guam in early August my crew was fortunate enough to be assigned to some air conditioned quarters in one of the nicer barracks, so it was better than it could have been. All the raids then were 30-plane raids at night in South Vietnam, and there were only about 25,000 'friendlies' (U.S. Forces) in South Vietnam at the time, so our instructions were to not bail out over land but to get over water if we had problems. That was before crews were issued sidearms or Chap kits, but we were permitted to carry your own sidearm if we had or could get one. Since most of the crews considered the raids in South Vietnam to be "milk runs" most people did not bother to try to acquire one. I thought differently, and carried a bandolier with 100 rounds of semi-armor piercing ammo, which was all I could get that met the Geneva Convention. I bought the ammunition at the Base Exchange at Andersen, and fabricated a holster for my gun (a Model 27 S&W .357) I thought I would survive a bailout, and put it on sometime prior to landfall on every mission in 1965. I don't think any of the rest of my crew at that time carried one, though.

During that summer and fall of 1965 most of the raids were at night. Since there were so many crews, and newer quarters had not yet been built, only about half of them were air conditioned. Trying to get crew rest during the day before a flight at Guam was pretty hard without air conditioning. Sometimes it was necessary for the flight crews with air conditioning who were not scheduled to fly to give up their quarters for a few hours to the crews flying that night who did not have air conditioning. There was generally short notice for the crew so displaced, and there was some personal friction on occasions during that exercise, but crews not flying respected the need for rest, so we put up with it. The crews who were flying were gone for briefing by 1800 anyway, so it amounted to giving up your bed for the afternoon, and since you were not flying that time could be spent at Tarague Beach, or if you could get transportation, at the Navy BX at their sub base on Guam. They had a much larger BX than Andersen did. Another favorite pastime for us lonely or homesick Crewdogs was to go to the passenger terminal on the Andersen flight line and peruse (ogle) the passengers who came off the planes into the terminal during refueling stops. Just seeing a pretty (they got prettier as the year wore on) girl after weeks of looking at fellow crew members was a welcome break in the monotony, and the food in the snack bar there was not too bad either, as I recall. Our quarters included a refrigerator for each crew, and with a bit of ingenuity and some trips to the commissary, it was not necessary to spend much time in the chow halls or Officer's club unless that was where you preferred to eat. My crew managed to obtain a grill of some kind, and we kept soft drinks and sandwich and other makings in the fridge as well.

Fatigue on those long flights was more of a problem than anything else. Our return flights to Guam were flown at 42, 43, and 44,000 feet for lead, two, and three respectively and we were supposed to maintain a two mile in trail separation. I remember thinking how stupid it was to try to maintain perfect separation if you were number three, since about

40 minutes out of Guam we were supposed to slow from 440 knots true to 400 to increase separation for landing to eight and 16 miles respectively. 400 knots true at 44,000 feet is not fun, even at light weight near the end of the missions. It is uncomfortably close to the slow side of "coffin corner", a term familiar to B-47 and other high altitude flight crew, which means the plane is close to stall speed. An inadvertent stall in a B-52 anytime is decidedly unpleasant. My crew generally started to increase separation early by slowing to something safer, like 420-430 KTAS.

Maintaining oxygen "immediately available" or actually being on oxygen was pretty tiring on those early morning post strike cruises back to Guam, and it was easy to nod off. I was not flying that night, but I later heard a story of one entire crew that went to sleep. They were flying number two in the cell at the time. That particular crew started to drift way out of position so number three notified lead, who attempted to contact them. He couldn't get anybody on the radio, so they kept calling and finally got the gunner to respond. He couldn't raise anybody up front so number three aircraft got out of the way and the number two gunner moved his turret to one side and shot a short burst. After the third burst the navigator woke up and ran upstairs to wake up the pilots, both sound asleep. I heard it was a stan-eval crew, but they were usually cell lead so that might not be true. I have a confession about going to sleep myself, in 1969 while watching my own bombs drop no less, so I won't pass too harsh judgment on a tired fellow crew member. But then that's another story.

At the time of the early Arc Light sorties it was considered normal for flight crew to stay strapped into the seats and wear the helmets all the time, at least for the pilots. I know the navigators and EW generally wore headsets much of the time, but since pilot's oxygen requirements were more stringent they were pretty much expected to tough it out. That was fine for a normal mission, as the combination of air refueling, low level, and transition training required everybody to be strapped into the seats anyway during those activities, and for a seven to nine hour training mission a couple times a week. At most it was not all that tough. "You have to be tough to fly the heavies", right? Only on very long missions like Chrome Dome was it accepted that during non-critical phases of flight (high altitude cruise, not in close formation) the ejection seat safing pins could be installed and a crew member could slide out of his parachute and harness to work with only a seat belt. Before the integrated harness became standard there was a separate harness and pilots were expected to wear those as well as the seat belt, though just sliding out of the parachute was still a big help in reducing fatigue. He still had to comply with oxygen requirements with his helmet and mask, but at normal enroute altitudes that was not difficult except for the pilots above FL 410 homebound.

I did not smoke, but some of my crew did and so on long flights at night I would often go to and stay on 100% oxygen to avoid the stink. After several incidents and growing fatigue with the number and length of the Arc Light missions it was finally deemed to be acceptable practice to go to the more relaxed seat and harness configuration. I don't remember exactly when it became official policy. I seem to remember that some senior crew (not mine) put the proposal to the staff and they issued the written approval so they could take the credit. Credit taking by staff officers was already a polished art in those days, but the Crewdogs were generally the ones doing the work and making the suggestions. I am always reminded of "the Not Invented Here syndrome" - a basic military axiom that to get anything changed it is first necessary to convince those with the power to implement a change that it is not only a good idea, it is really their idea. You also should show them how to be able to take credit for it, though they are usually pretty good at that. CEG, later called CEVG, not to mention the numbered Air Force and SAC IG, were, in my experience, the masters in practice of this axiom. That never changed during my career, both in and out of SAC.

During that summer and fall of 1965 tactics and bombing gradually evolved. Early raids were mostly individual synchronous bombing runs using radar offset on a geographic feature, but some of them were conducted as raids in WW2 had been. The lead airplane in each cell did the navigation and bomb run, with numbers two and three on the right and left wing respectively. Some of them were conducted using radar offset on a beacon, with each airplane synchronous bombing individually and only going up to release on a wing in case of radar failure. There was one time a crew bombed a beacon direct by mistake, rather traumatic for the poor helicopter carrying it. There were only a few missions early on that were not 30-plane raids. All required prestrike refueling on Busy Rooster A/R track, and I think our offload was 93K per plane. Tankers came from Kadena or CCK, and some missions had our tanker launched from Guam and recovered otherwise, but I don't remember the tanker operations very well. My AC was a good stick, always got his fuel in good time, and for the most part my crew (except for that fourth mission) got bombs on target every time.

We would occasionally get a "hanger", or indications of a hanger, and have to do an in-flight bomb bay check prior to landing. That did not please the navigator at all, who had to put on a chute sometime before landing, and go through the wheel well to visually check the bomb bay. Had there actually been a hanger he could determine whether the arming wires had been pulled or if it was simply a rack malfunction and the bomb was reasonably secure. Crew action afterwards was dictated by the condition of the bomb. Some fairly hairy incidents occurred during the war as a result of hangers, but my crew never had one. I heard of at least one incident when the pilots forgot there was a man checking the bomb bay and lowered the gear before his return. This undoubtedly caused a loss of harmony on that crew. I heard that some crews actually pulled the normal gear control circuit breakers for the forward gear when they had to do an inflight bomb bay check, but I didn't think that was a very good idea. I wanted a continuous gear up signal to reach those gears if a man was in transit near them and my understanding of the circuitry was that loss of the normal gear control circuit might enable corrosion or a stray electrical signal (like lightning or static) to cause the gear to cycle without command. I thought my crew was smart enough to remember when somebody was near them. External hanger indications were almost invariably due to water freezing or corrosion, and went out prior to landing.

I do remember one memorable mission in which the lead airplane led us into the top of a thunderstorm. Lead of each cell had the responsibility for deciding whether to continue or abort a bomb run for thunderstorms, and unless the mission was a "press-on" or wartime nuclear mission, we were supposed to observe peacetime rules regarding thunderstorm penetration. Anyway, our altitude changed over 1,000 feet (almost uncontrollably) during the last 120 seconds of the bomb run. After we got back to Guam my AC, who was about six feet four and 250 pounds, sought out the lead AC and I overheard a heated heart-to-heart discussion. That never happened again.

There was one incident that occurred, I think on our first mission, on 4 July, 1965, that is worth mentioning. The airplanes had been sitting loaded with bombs for some time, since it had been 16 days since the first mission. During the first plane's takeoff he lost several external bombs, later blamed on corrosion in the MER racks wiring, caused by the rain and salt air on Guam. The bombs came off during the takeoff roll undetected by the crew, and slid down the runway in very close formation. The rest of us had to wait till the ground guys cleared the bombs, which stopped scattered all over the runway and adjacent areas. They spent about 30 frantic minutes using whatever they could find to drag the bombs clear of the runway. The M-117s had come off the airplane safe - their arming wires still secure to the bomb, so they were in little danger of going off. I remember my grandfather's admonition that if you work with poop long enough you will get dirty and it was still pretty dicey for the guys moving them. After they were clear, each plane was watched closely during takeoff, but as I recall only the lead plane had any problem. It was fortunate that there was still enough daylight to see, as otherwise subsequent airplanes might have had a close encounter of the worst kind. Fortunately too, we did not launch MITO, but there was one minute spacing between planes in each cell, and about 15 minutes between cells.

Finally the end of our tour came and the unit began leaving for Carswell, as the Mather and Columbus crews began to replace the Carswell crews. My crew rotated home among the last, of course, in early December of 1965 after we had flown 26 missions between 4 July and early December. It was our fifth Pacific crossing in the back of a tanker since February of that year. All the flights were made with a full complement of some 55 crew or staff or maintenance members, and whatever baggage or tools or loot we thought it necessary to take. At the time, there was a KC-97 (called the Yokota Flyer) which made round trips regularly between Andersen AFB and Yokota AFB in Japan, and it was possible to buy a new Honda 90 motorcycle in Japan for $90 and a tip to the KC-97 crew chief. They would bring one back to Guam and you could use it for transportation around the island. Since Guam had little topsoil, and the coral under whatever there was got really slick when it rained (which it did all the time at Guam), there were more than a few accidents. I recall at least two crew members who got badly hurt during that time. Nevertheless, if you drained the fuel and oil, you could take the motorcycle back to Carswell on the tanker. They, as well as the Hibachi pots popular at the time, made crew baggage pretty heavy. I took home at least two full sets of Noritake China, which cost $39 for a white and around $49 for the average pattern 12 place setting including over 100 pieces of china. Those boxes of china were heavy. Everybody took his duty free gallon of booze, of course, which we bought at the Class 6 store in the passenger terminal at Andersen. I remember Chivas Regal was $3.20 per fifth, there was a 12-year-old Ballantine Scotch which cost $3.60, there was a Scotch I didn't think I could afford for $5.15, and fine liquors like B&B and Drambouie cost around $3.

But anyway when we redeployed to Carswell on the tankers, they were really loaded. I sat in back for two tanker takeoffs on RWY 6R on Guam, which involved taxiing to the end of the takeoff overrun, the engines set to max-power-wet before releasing brakes, and I remember watching 65 seconds elapse on my new SIEKO from brake release to liftoff, rotating in the far overrun. The plane used some of the 600 feet above water off the end of the cliff to retract flaps and get to climb speed. I am glad I didn't know how heavy the plane was, but I suspect nobody really wanted to know. Some time later, I think it was at CCK, a tanker so heavily loaded was lost after an engine failure and they began to actually weigh luggage.

I don't remember much about that final trip home in 1965, except that when I did get back to Carswell and cleared customs I did not have a place to stay. During that November, my landlord (the retired officer I mentioned a few paragraphs ago) had decided he wanted his house back and that full story doesn't belong here. Suffice it to say, I lived in a motel for a week until I found another house to rent in Ft. Worth. Some of the earlier returning crews had helped my wife move our stuff into one of their garages, and I borrowed my AC's pickup to move to the new house during the time I had off before we resumed a more normal peacetime posture.

Sometime in that period, perhaps earlier, I don't know for sure, the decision was made to upgrade the D-model fleet for the big belly mod. After the crews who replaced us in December of 1965 finished their tour, the D-model units took up the bombing and I flew no more Arc Light until I transferred to Grand Forks AFB in 1968. Later I upgraded to AC and after three solo missions with them, took my first crew Arc Light in 1969 as GRA R-18, an RTU crew.

Cu Chi Vietnam 1967 - Jim Bales and Larry Moeller.

First SAM
Larry Moeller

"Pepperbasket!"

The immediate Surface-to-Air-Missile (SAM) threat codeword shatters our bomb run radio silence.

No translation needed: ABORT & get the hell away! "Dive - dive - dive" as the submariners might say in a WWII flick.

Diving, rather - simply falling out of the sky ASAP, is the only realistic survival response a B-52D has available. This is also known as the "falling leaf maneuver."

BUT…."it's too late!" Our bomb doors are open and we'ved just pickled off the first of 108, 500-pounders.

"Radar acquisition 12 o'clock" called EW Jim Bales on intercom.

"Oh, Shit", says your homunculus. Though we can hear the SAM chirping via the APR-25 radar warning monitor; it is naught but a niggling background noise as we focus on the "emergency at hand".

"Jamming", reports Jim, then "Chaff", and again, "Jamming."

"Missile launch" says Jim, signaling that a BG-06 missile guidance signal had activated the launch light. "Jamming"

"Christ", you think. "That's not right! They can't turn it on and launch that quick." The SAM mobile launch sites are supposed to have to tune up their radars prior to launch, and that's when the Electronic Warfare (EW) F-105 Wild Weasels finds 'em, and we'd have been aborted.

There's also supposed to be about 30 seconds from initial radar target acquisition to launch, with a second radar antenna that provided a BG-06 guidance signal to steer the missile and trigger detonation altitude. Theory was - that gave us time to take evasive action. "Damn …. how long has it been, already?"

Pepperbasket!" "Pepperbasket!" urgently says the Wild Weasel on the radio. "This is serious", you think. But you don't know to be really scared, because this has never happened to a B-52.

"Jamming", reported Jim, then "Chaff", and again, "Jamming."

It was April of 1967. We are flying right on the southern edge of the DMZ, somewhere over Khe Sahn, as the lead ship in a two cell formation of six B-52Ds, on a mid-morning strike at 29,000' or thereabouts. We were supporting "Operation Manhattan". SAM attacks by mobile launch units were real threats anywhere near that that airspace.

"Pepperbasket!"

You think, "This isn't supposed to happen". It's *"Twelve O'clock High"* and YOU are there. Buffs are big fat slow targets.

The mission briefing "Intel" that morning had assured us that the Combat Air Patrol (MiG CAP) and Wild Weasel electronic surveillance fighters would clear the DMZ target area. Standard procedure was for eight to 12 fighters preceded the B-52s into a hostile zone and remain until they departed. Usually, the Weasel pilots could locate a SAM site by getting the North Vietnamese Army (NVA) to fire a SAM at them, thus revealing their position.

This was supposed to be a "milk run". It was a Clear and Visibility Unlimited (CAVU) day. We were a lead crew from the 744th Bomb Squadron out of Amarillo, deployed to Guam on Jan 5, 1966. The Aircraft Commander (AC) is Captain Lowell "K" Hanson, a Yale grad and Regular Air Force "lifer". (K eventually retired as an Lt Col).

Hours of SAC training and memory-banked auto-psychomotor reactions plays in your head like movies.

"Radar, report bombs away" ... coolly calls "K", as he held the BUFF steady on course and altitude, while I tweaked the throttles to maintain precise bomb run airspeed.

I felt with my right hand for the reassurance of the ejection handle.

Christmas Day, 1967. We were having a shared Christmas Dinner at the Amarillo AFB Alert Shack with our wives, (the good-old-days), when the Wing Commander called us into an emergency briefing and announced the 461st Bomb Wing was being deploying to Anderson AFB in one week. I was a fresh copilot, Vance Class 66-F, just six months out of Castle, with only two months of Alert Crew experience. I'd thought choosing a B-52 assignment would guarantee a slot as airline pilot in four years, home-base stability, and reasonable safety compared to the only alternative, an F-4 back-seater job with Pacific Air Forces (PACAF). USAF jet pilot work in those days was still dangerous business.

It was hard news. Bunnie, my wife of one year was so devastated and fearful that she refused to come to the base to see me off to the war on January 5th when we'd clambered onto a KC-135 carrying a spare B-52 wing flap and 70 other guys. Only six weeks earlier a KC-135 had gone in on the Amarillo runway with no survivors. This day's takeoff was a near miss. The 135 lost water injection and I saw telephone poles going by as the pilot yanked it off at the end of the overrun. Good thing Bunnie wasn't there to watch. So it goes.

"Report doors shut." calls the A/C

Slow Time: Seconds pass like hours.

We had a bomb run to complete and no chance to abort. 500 hours of flying experience plays in your head in silent, vivid reruns.

Your mind races at warp speed. You rehearse your imminent part in the silly Falling Leaf Maneuver that we'd recently begun practicing from 42,000' on the long return trips to Anderson. (As if anybody REALLY believed that would work.) But nobody else hears. Nobody else sees your private movies.

You write that letter to your wife.

"Pepperbasket!"

Each of us is at once multitasking, supremely busy doing our own fluid part in that choreography inherent in superbly trained aircrews, parts of a single perfect machine, while playing one's own private theater in tandem, -silent, -fast-forward.

Each of us is thinking "How long will it take for those 108 bombs to fall out?" How many seconds to close those huge radar-target bomb doors?" "Oh, my God .. we've had so many hung bombs.... PLEASE, let this be a clean drop."

"'Jeez, how fast do those SAM's get up to us?" "How long has it been already...?"

"Jamming" says Jim, fairly matter-of-factly. "Chaff." It would be later in the air war that we'd learn chaff was irrelevant.

"Bombs away" says Radar Nav "Dutch" Vanstockum - and he hits the Emergency Armed Released (EAR) switch to ensure a clean bomb bay.

"Doors Shut" gushes Dutch, as you feel and hear that "snap-clunk" shudder thru the airframe.

I pull off the power. "K" twists the yoke and heaves the groaning machine into about a 50-degree left bank and lets the nose fall into a dive. We go as knife-edged as a BUFF can get.

"Speed is life. Altitude is life insurance." Or not?

"Missile detonation!" "Above and 3 o'clock - about where we would have been", reports Gunner Lanny Passmore.

And then we are out of there.

Falling, diving, banking as fast and hard as any BUFF can. The falling-leaf maneuver originated during WWI as a flight training exercise. Pilots intentionally stalled the aircraft and forced a series of incipient spins to the right and left. The aircraft descends as it rocks back and forth, much as a leaf does falling to the ground. I guess it worked.

You manage a glance at your Seiko, one of the ones we all bought at the BX on day two of our deployment, and synchronized before every launch. About three minutes have passed.

Long Time: The recovery hours back to U-Tapao pass like months. Navigator Brad Shelton has figured out where we are and plotted a course out to sea. Nobody talks. You open your in-flight box lunch and discover God as a hard-boiled egg. You munch on cubes of roast beef and the baggie of pineapple and mango you bought outside of the U-Tapao gate that morning.

At debriefing, hours later, mentally exhausted and sweaty in our flight suits, we learn that "Intelligence" now says Russians have apparently upgraded the SAM launch computer systems to allow actual missile launches before turning on and "tuning up" the radars.

We were the first target. The B-52 air war had just become a lot more dangerous.

And you wonder - "Did they know we were coming? Or was it just a coincidence? But why did they ignore the Weasels if it was only coincidence?"

May of 1967 - We are in Vietnam, on the ground, during Operation Manhattan in Cu Chi.

As flight leaders, our "50-Foot Center-of-Error" bombing accuracy scores "won" us a free trip to Vietnam with the "Big Red

One" to observe Arc Light bombing effectiveness. A grand adventure. We looked in water-filled bomb craters, visited the Cu Chi tunnel-rat school. We came under VC attack in a "hot" chopper landing zone (LZ) and experienced a harrowing extraction through treetops aboard an overloaded Huey. Now THOSE pilots had the real balls.

We learned that one of the Army goals for the B-52 mission was to kill trees to create Landing Zones for choppers. We spent an interminable night on cots in a shell-pocked concrete room in a former Michelin Rubber Plantation, then HQ for the 9th Infantry. The walls had lots of brown blood splatters. Mortars and small arms punctuated the darkness. We stayed one night at the Eden Roc Hotel in Saigon, watching ground and aerial fireworks from the rooftop. Later, as we walked around to see the sights, a grenade-laden Viet Cong (VC) was spotted on the street, forcing us to hole up in the USO behind sandbag barricades for about four hours.

Army Officers told us that the ritual of defining B-52 targets required clearances from local "tribal chiefs" who had to have notice enough to evacuate good-guy civilians well in advance of any strikes. VC alliances, graft, corruption and politics often determined the real effectiveness of B-52 strikes. I believe the D.C. politicians called that "Rules of Engagement".

And the SAM question returns. "Did they know we were coming?"

We talked to lots of "grunts" who lavished the B-52s with praise for the salvation that the fearsome strikes brought to guys on the lines. Even today, 40 years later, I still hear that awestruck thanks from guys who were there on the ground. It still brings a lump to my throat. That can never be forgotten.

I will also never forget bombing "MILKY" at Dong Ha one very dark night.

Callsign MILKY was a COMBAT SKYSPOT (CSS) "ground directed bombing" radar site at a Marine base near Dong Ha. Sources say that Skyspot could deliver bombs on targets within a few hundred feet or less. CSS sites dropped somewhere in excess of 75% of all B52 strikes. In early 1967, we only knew "Milky" as Dong Ha. It may actually have been at Phu Bai, on Monkey Mountain.

We were flying lead in a three-ship trail formation, in the final stages of a routine bomb run. On the radio you could hear muffled machine gun fire and explosions. That was extremely unusual. You knew the VC were damn close to the radar site.

The radar operator was disciplined, cool, and matter-of-fact. It was clear to us, given our briefed backup targeting coordinates being monitored by Dutch, that MILKY was shifting the target to a point closer the base. Maybe, right on the base? He said they were being over-run.

In the final moments of his direction, the sounds of gunfire and machine guns became terribly loud. You could only guess that the doors to the mostly underground radar "shack" had been breached. He called for bomb release. There was a loud explosion and the radio went silent.

Long time. We were flyboys then, and young....

We flew 50 missions "on the record" and re-deployed to Amarillo in July of 1967. There were another dozen "off-the-record" missions over Cambodia that don't show up on my "Master Flight Record - Pilot". Bombing the Ho Chi Minh Trail in a neutral country was not legal, so we didn't record it. I was sent to survival school in September of '67 (go figure) and while in Spokane we learned the 461st - 744th was being disbanded. Six months later I was in G-models at Beale. Avoiding going back to Vietnam by volunteering for B-66's, I got a hiring bid from United Airlines in October of 1969.

I flew Vietnam support a couple of more years with a C-124 AFRES outfit out of McClellan, teaching school nearby, while waiting for a United school date that eventually succumbed to the 1970's recession. So, I left the USAF behind to earn a Ph.D. in education. I can retire in 2009 with a good pension. Jim Bales left to become an architect in Florida. I think Dutch and Brad stayed in USAF to retirement. Lanny went home to Washington State, as I recall.

In the haunts of my mind, I will always be a 25-year-old bomber pilot, and yes, I still have the Seiko wristwatch, but the gaggle light is long gone.

The five that flew Linebacker II, P - Glenn Russell, CP - Les Spencer, RN - Doug Jamieson, EWO- R. J. Smith, and Gunner - Mac O'Quinn.

Don't Shoot Three - it's Two
Glenn Russell

I was crew commander on an Arc Light deployed crew from Fairchild AFB, Washington, with my copilot Les Spencer; RN, Doug Jamieson; EWO, R. J. Smith; and Gunner, Mac O'Quinn. At the time this event took place we were without a regular navigator on our crew, for reasons that I will not go into at this time. Since we were without a regular navigator, we had substitute navs on all our missions during Linebacker II and after all these years I don't recall the name of the navigator that flew with us on each mission. During one of those Linebacker II missions flown, we were flying a B-52D in the number two position of a three-ship cell out of U-Tapao. The target that night was in the Hanoi/Hai Phong area.

We all knew the importance of the missions and the targets we were assigned during Linebacker II, and during the intelligence phase of the briefings we were informed of the fighter and SAM threats that lay ahead of us that evening. Our takeoff and climb out from U-T were normal but shortly after we leveled off we noted that the needle for the oil pressure gauge for engine #5 was sitting at zero. We swapped oil pressure around to see if it was just the gauge that had gone bad, but it turned out that it wasn't. Considering the mission ahead of us I didn't think that it would be a good idea to shut the engine down since we were going into a high threat area. We needed all the power we could get that night. I thought that it might make the engine run longer if I reduced the power to idle, but we were too high and heavy to maintain airspeed if we did. As a result, we lost about 20 knots airspeed and so we decided that it was necessary to bring the power back up on #5. By that time we were getting close to the IP and needed to get set up for the bomb run.

The bomb run was very busy and required a lot of maneuvering, dodging the Surface-to-Air Missiles (SAMs) that were being fired at our cell. After the bomb run was completed we headed out over the Gulf of Tonkin and things started to settle down a little and we knew that we really needed to get back into our appointed position in the cell. Even though all the crews were well trained and determined to drop the bombs on the assigned targets, it was not uncommon for cell integrity to be broken in the process of avoiding the SAMs after bomb release. The Radar Nav said that the cell's lead aircraft appeared to be in our ten o'clock position and so I started to maneuver into a position to rejoin it. At that point, the EWO called a lock-on at ten o'clock, the Radar was reporting chaff, and I saw that the aircraft ahead of us was dispensing flares.

I quickly realized that was not a good position to be in and made a turn to get away from what I figured must be our number three who had somehow gotten ahead of us. At the same time I got on the cell frequency and said something like "Don't shoot three - this is two." After all the confusion settled down we were able to get back into position and return to U-Tapao. Our #5 engine completely failed about 30 minutes later, but at least it did not do that during the bomb run or post target turn.

After we landed and debriefed we checked with number three and found out that the gunner thought we were an enemy aircraft and had fired approximately 150 rounds of 50 caliber armor-piercing incendiary rounds at us. Since the B-52 gunner only had radar and did not use optical sights at night, it must have been difficult to tell the difference in the orange glow of his fire control scope between a friendly bomber and an enemy fighter. Luckily, he had missed. Mac O'Quinn believes that we must have been outside of his cone of fire and that was why he missed.

This was a very experienced crew and had been together since the start of the Linebacker Operation. We all had previous SEA experience. Les Spencer, Doug Jamieson, and I had each spent a year in Vietnam as crewmembers on other airplanes. R. J. Smith and Mac O'Quinn had racked up lots of Arc Light experience. In fact, R. J. flew more Arc Light missions than any other B-52 crewmember, with a total of 506.

We flew six missions together during Linebacker II, and I think we may have come closer to getting shot down on that mission by a fellow B-52, that any of the other missions we flew. I thank my lucky stars for having a good crew and someone looking out for us.

A Night on the Town
Wade Robert

I guess when your hair is gray and you look back on flying the B-52D, you realize that of all the missions you flew some of them must have defined the B-52 experience for you. For me, it was those nights I spent over Hanoi, North Vietnam and not my more than fair share of the other 126,615 Arc Light sorties that were flown during the B-52's Vietnam War. The problem is that when your hair turns gray so does your memory. I remember the stories of the crews who got shot down and didn't come back. I never tire of reading about them and thinking of old friends. I have also been to The Museum of Military History in Hanoi and I have seen the collage of wreckage in the courtyard that incudes a B-52D vertical stabilizer and I can recite the unit numbers of the most vicious surface-to-air missile (SAM) sites, especially VN-549 that is memorialized there. I know I flew Linebacker II missions over those 11 days between December 18th and 29th of 1972, that is still called the Eleven Day War by crewmembers. Nevertheless, the missions I flew are still vivid and arguably tame compared to many of the heroic Linebacker II stories I am familiar with. So with this in mind, I'll tell you my plain story as I remember it 36 years later about my first Linebacker II mission. It is probably fairly typical of those crews who just did the job and made it by hook or by crook and returned to fly another day.

The evening of December 18, 1972, while Hanoi was getting its first taste of the B-52, March AFB crew number S-22 showed up on the flying schedule for the next day. We would be flying on the second day of the offensive. We had been to the World War II-like massive crew briefing two evenings before the first sorties, where the senior officer briefer dramatically opened a big curtain on the stage of the briefing room at the Anderson AFB Arc Light Center after saying something right out of the movie *"12 O'clock High"* like "Gentlemen, tonight your target is ...HANOI!" The curtains flew open and we saw the big map of North Vietnam that had a big red bulls-eye with HANOI written across it. What a surprise! The briefing room remained totally silent and I am sure cheering was expected similar to the reaction of General Savage saying "BERLIN," but on balance the crews sat there with very serious looks on their faces and mentally screaming "Oh, Shit!"

By the time we arrived at our real mission briefing on December 19th everyone had already been told what the trip downtown was like. We had seen the squadron commander and chaplain visit some of the wives who were on Guam at the time. I knew it would be a hard mission – one that I had thought about for four years. So we became Orange 2 and we were going to the Hanoi Rail Yards that night with a load of 108 Mark-82 500 pound bombs and all of our fly-no-die Buddhas and other crew good luck charms.

I remember the trip to the aircraft that afternoon in the big Blue Bird school bus and even though I had watched the hours-long launch the day before, I was awed by the shear mass of the assembled strike force on the Anderson AFB, Guam ramp. My copilot, Dan Kinneburg, and I did a normal pre-flight and soon our engine start time came and the workhorse Pratt and Whitney J-57 engines started their familiar whine. About that time we became our cell (formation) leader since one of the already airborne B-52s had aborted the mission because of engine losses (yes, losses, and if I remember correctly that bomber later landed with four engines out on one side). Once the bomber operations controller, Charlie, cleared us to taxi our cell entered the long line of bombers awaiting the ever thrilling takeoff down the bowed Anderson runway. There were no taxi or take-off clearances from Charlie that day; it was just a moving stream of airplanes to the runway. As I pushed-up the throttles and the B-52 crawled-up the slightly uphill curved runway hammerhead there was only a single call to be made that day - the aircraft commander's name and cell position - "Robert, Orange One". We pushed-up throttles and the monster started its labored trip down the runway, followed by the roar and kick of the water thrust augmentation that the B-52D used to assist the take-off. The airspeed rose rapidly and committed us to the downhill portion and the standard Guam takeoff.

The uphill portion of the take-off roll followed where the airspeed and my heart lingered below takeoff speed until finally moving and slowly reaching the speed that allowed us to get airborne and watch the cliff disappear and my copilot's hand come down from the bomb jettison switch. The hundreds of feet of altitude we got instantly from the cliff at the end of the runway always made us feel like we could get rid of the bombs should the B-52 fail to gain altitude at the end of the runway. I was never sure that it would work, but it made us all feel better. After a big sweeping left turn to the west, Orange cell began the long voyage to the Glasstop refueling area near the Philippine Islands. Guam disappeared rapidly behind us.

The flight to the Glasstop was the same as it always was, our navigator Ken Stegemiller was navigating to the refueling area and Danny Hertzner our gunner was ranging and azimuth checking his guns. Neither I nor my copilot was drowsy that evening like we normally were. We just kept chasing the lowering sun in the West. Soon, radio calls were being made to the KC-135 tankers by our radar navigator (RN) Dave Stuart as we all prepared to refuel. It was then that we found out that the tanker force from Kadena AFB in Okinawa had been scheduled (fragged) an hour and 20 minutes early. The radar navigator went out in the B-52's radar beacon mode to find our KC-135 tanker, somewhere, and the entire scope was flooded with tanker radar beacon returns. All he had to do was to find our specific tanker formation somewhere in the mass of beacons. I then heard the most brilliant radio call from Dave. He asked for all, yes all, the tankers (and there were hundreds of them) to go standby on their beacons except our tankers. As if by magic there was a single beacon on the scope! He soon gave me heading to our tankers and in what seemed like an instant I was echeloning Orange cell and clearing them to the KC-135s.

The refueling was difficult on that mission since there were our three B-52Ds refueling behind four KC-135 tankers that were echeloned to the right of their leader. I started refueling on the number two tanker (leaving the number one tanker waiting), got part of my fuel offload while Orange 2 got part of his offload on the third tanker and Orange 3 waited his turn. Each bomber then shuffled to the left placing me behind the first tanker where I finished my offload of 120,000 pounds (about 18,000 gallons) of JP-4 aviation fuel, and so on in order. If anyone had screwed-up the refueling the resulting varied offloads would have seriously affected our mission and probably resulted in someone diverting to U-Tapao Royal Thai Navy Air Field to land after the mission. That was my first of many heavy sighs of relief that day.

I'm not sure what I did as we continued west, but I remember that somehow our entire wave of bombers had gotten our gas and the formation was intact. I probably nibbled at my cold greasy chicken and smoked cigarettes while I went over procedures in my head. I have never since anticipated anything in life as I did my pending arrival over Hanoi. I had seen SAMs by that time, but many of the crews had never seen one. There was very little chatter on the intercom and there were long periods of silence on the radios. We entered the timing boxes on time and had very little adjustment to make and left for a point that was called PL or Papa Lima by the crews which was the entry into South Vietnam. We proceeded across South Vietnam looking down on the always present isolated fires and occasional low level tracer activity from in country fire fights and then entered Laos and flew northwest almost to the Chinese border. Our F-4 MiG protection (MigCap) was coming off their tankers and our F-105 Wild Weasel SAM suppressors were near our formation. The F-4s, call sign Merkel, joined behind Orange cell maintaining high speed in orbits in case they needed to protect us from MiG fighters. Pre-target checklists were run by the navigators and pilots.

When we were almost to the Chinese border there was a call from downstairs to turn southeast. We had turned southeast down the famous mountain called Thud Ridge that pointed to Hanoi and we were truly on our way downtown. We had checked-in with Red Crown, the Navy Ship offshore that controlled the airspace over North Vietnam and told them we were "...in with three and on time." I remember thinking it strange that I was really going down Thud Ridge in a B-52D. I had read Jack Broughton's book *"Thud Ridge: F105 Thunderchief Missions Over Vietnam"* before I went to pilot training and Thud Ridge had a reverent place in my mind. I also remember saying that this was one of the same routes the B-52s had taken the night before and that the SAM operators were probably looking at their watches saying, "Hmmm, it's about time for Orange cell to get here." Even on day two we crewmembers knew this standard and repetitive routing was wrong even though the mission planners at Strategic Air Command headquarters did not.

It was dark on the ground as far as the eye could see and there was a low lying layer of clouds and haze. It was clear at our altitude. Suddenly there were lights that flashed bright about two miles or so at 11 o'clock high and I immediately called to our cap Merkel to point out what looked like landing lights flashed on and off. The response I got was "Roger, Orange, that's Merkel 4 trolling for SAMs." I was dumbfounded, but I had always been dumbfounded by the things fighter pilots did. I glanced down at the green terrain avoidance radar scope and saw a small blip with a bright sideways "V" streaming from it. The F-4 chaff droppers had started their run to confuse the North Vietnamese gunners. For a long time it was silent, but then the first "SAM launch! Sam launch Hanoi!" calls started. Far ahead I could see the strings of bright light rising upward and I knew that B-52 cells ahead of us on their bomb run were being attacked by SAMs. One could almost follow their route through Hanoi by looking at the SAM exhausts and warhead detonations. Hanoi had emerged from darkness into an eerie red, yellow and orange glow that showed through the cloud cover over the city as the bombs from the lead cells reached their targets. Our electronic warfare officer (EWO) Bob Brenizer began to call that SAM acquisition radars where looking at us and told me that we had an F-4 behind us. We were getting close.

Shortly after the EWO had notified us of the SAM radars looking at us the copilot said, "Two SAMS 11:30" and I started maneuvering the airplane in a 20 degree, 40 degree corkscrew which was the only approved and totally ineffective maneuver that B-52s were to use to avoid SAMs. Like the copilot, I had seen the bright flashes under the cloud deck as the SAM left its launcher, and then the clear bright yellow exhausts as they cleared the clouds then followed by a flash from each missile as it ignited its second stage. What seemed like a long time was really only seconds before the SAMs were close to us. The actual cockpit and radio conversation went like this after the initial alert from the copilot:

Copilot (CP):   "I gottum – they are in front of us."

Electronic Warfare Officer (EWO): "SAM uplink!"

Note: The SAM uplink is the main guidance signal to a SAM from the ground controller after it is fired. The EWO sees this signal and tries to jam this guidance signal and cause the missile to miss the aircraft. I'm sure he caused many to miss.

CP: "Detonation one."

CP: "Some more coming up."

CP: "Two more just came up"

Radar Navigator (RN): "Jesus Christ!"

CP: "Four more"

CP: "Clear at 10 o'clock."

CP: "Eight in the air but we're clear."

The navigator gave me a turn and we were initial point (IP) inbound – the official start of the bomb run:

Navigator (N):"Pilot turn now 242 degrees."

Pilot (P): "Orange, turn to briefed."

Note: We pre-briefed headings so as not to broadcast the actual heading.

GUARD FREQUENCY (radio call on all frequencies): "Multiple SAM launches, Hanoi!"

Note: I never knew where this voice came from broadcasting over the GUARD FREQUENCY but I'm sure now it was an SR-71 orbiting over Hanoi. We would see a launch and then the broadcast would come so we were probably seeing the same thing.

CP: "Detonations ahead and above us."

RN: "They're getting closer."

CP: "Two more SAMs came up."

GUARD FREQUENCY: "SAM launch! SAM launch, Hanoi!"

CP: "Four more just came up."

GUARD FREQUENCY: "SAM launch. SAM launch Hanoi!"

CP: "Look at 9 o'clock."

RN: "PDI is good, IP inbound, in bomb."

Note. The radar navigator was telling us that I could follow the position direction indicator (PDI), a pilot's instrument that is slaved to the bomber radar and followed to the target by the pilot by centering a needle. He was also saying that his system was ready to drop bombs."

CP: "OK, 3 more at 12 o'clock."

Orange 2:   "SAM 11:00 low."

RN: "Are they close?"

CP: "This one's passing right."

RN: "PDI going left"

N: "180 seconds."

RN: "Center the PDI."

P: "Manuvering."

CP: "Seems clear."

Gunner (G): "SAM! 9 o'clock low!"

EWO: "SAM uplink, Orange."

CP: "Where is it?"

G: "Detonation 9 o'clock low."

At that point I remember being totally engaged without any thoughts other than keeping the B-52 on course to the target while avoiding any SAMS that I could. If the SAM was a bright glowing yellow ball on the windscreen and not moving around it usually meant that this missile was tracking right on you and likely to hit you. Against the proscribed maneuvering rules, I stood the B-52 on one wing with near 90 degrees of bank and missed a SAM that was glued to my windscreen. Then, amazingly, we got back on course. We were almost over the rail yards.

CP: "We've got AAA."

Note: These were the big 100mm anti-aircraft guns that were not very effective against the B-52.

N: "120 seconds."

RN: "PDI moving right."

Orange 2:   "SAM 11 o'clock."

EWO: "SAM uplink, SAM uplink!"

G: "SAM launch 8 o'clock low!"

G: "Two SAMs, 8 o'clock low!"

N: "80 TG."

Note: Time to go (TG)

N: "70 seconds."

N: "60 seconds."

RN: "Doors."

Note: The radar navigator opened the bomb doors. When the bomb doors opened we became a much bigger target as the doors acted as large radar reflectors.

Orange 2:   "SAM 12 o'clock."

CP: "SAM 12 o'clock, turn left!"

N: "30 TG, RCD connected, bombs are armed."

N: "20 TG."

CP: "SAM coming one to six!"

RN: "BOMBS AWAY!"

And then the yellow bomb release light flashed as the bombs fell from the aircraft with a momentary floating sensation as the old gal relieved herself of her deadly cargo. I prepared for the 60 degree bank post-target turn.

N: "1, 2, 3, EAR, 4, 5, 6, Turn!"

Note: EAR or emergency armed release was a switch the navigator used to hopefully make any bombs that had failed to leave the bomb racks release. It was a back-up to ensure all the bombs where gone.

GUARD FREQUENCY: SAM launch, SAM launch Haiphong!"

Right after that call we heard an emergency locator beacon start warbling over the Guard Frequency. The beacons were located on parachutes and activated when the parachute opened so clearly someone had bailed out. And then there was another, and another, and another, all warbling at the same time. It was chilling.

Immediately after the post-target turn I had a real feeling of relief. This was conditioning I had because of the numerous times I'd relaxed after a bomb release on much less threatening South Vietnam targets over the years. But the relief was very short lived. We were flying down a jet stream on our way to the target and our ground speed had been about 600 knots. At bomb release, we had broken right to a heading of about 270 degrees, head-on into the jet stream. Our ground speed dropped to about 350 knots and it seemed like it would take forever to cross the "fence," out of North Vietnam, especially with SAMs still flying at us. It started again:

GUARD FREQUENCY: "Sam launch! SAM launch Hanoi!"

P: "Can you see it?"

G: "SAM six low."

G: "Two SAMS."

Me: "Which way to maneuver?"

G: "Break left!"

Me: "Is it moving with us?"

G: "Yes!"

G: "Detonation six low."

EWO: "Uplinks again."

EWO: "Two uplinks"

P: "What's that bright light?"

GUARD FREQUENCY: Warbling parachute beacons.

GUARD FREQUENCY: More warbling beacons.

G: "It's a flare."

RN: "Oh, shit!"

GUARD FREQUENCY: More beacons.

P: "Someone took a hit."

We were getting really close to being out of North Vietnam and it had gotten really quiet after that round of SAMs and, despite our snail's pace, it looked promising. And then Orange Two and my gunner called more SAMS coming up at our tail. Fortunately they missed. I was really mad. The sneaky bad guys knew our route so well they had positioned SAMs along the egress route almost to the border. I guess it wasn't over until the fat lady sang, but that catchy phrase hadn't been created yet. Finally, the navigator said it was 30 miles to the fence and soon we were calling Red Crown telling them Orange was over the fence with three and we were on the way home. After a while I noticed an airplane flying alongside about a mile from our formation. At first I thought it was an F-4 going home and as it crossed our path and flew away we realized that it was probably a B-52 cell. Most likely they turned early on the way out and had cut us off. I never did find out his cell color, but it could have been a giant mess.

The airplane and radios became very quiet and we eventually coasted-out of South Vietnam. It was still dark and the weather had turned really foul. We were in the soup on instruments and bouncing around as we prepared to find our two post-target tankers. It was just what we needed, a night refueling in the weather and with three bombers refueling on two tankers. No gas, no Guam and an unscheduled landing in Okinawa, Japan.

We rendezvoused with the KC-135s while flying in and out of clouds and coping with some strong turbulence. The tanker boom operator found the refueling receptacle and we started taking on gas. Then came a pocket of fairly severe turbulence and I heard the boomer say "Orange One, breakaway, breakaway!" and out came the boom and the tanker started to climb. The normal breakaway procedure was to reduce power and descend 1,000 feet and clear the tanker as fast as possible. I'll never know why that boomer called the breakaway. We had been bouncing around together in the turbulence and from everything I could see there was no danger of the airplanes hitting each other. Since our flight was sharing tankers again we didn't have the time or gas to start over, let alone lose the tankers in the weather above us. Instead of descending as I was supposed to, I climbed with the tanker and when he leveled-out I was still behind and slightly below him. It was my tanker and he wasn't going to get away despite any procedure. Then I heard the boomer say "OK, if you're going to be like that, come on back in." We completed the refueling.

We were worn out and not looking forward to the long flight back to Guam. I imagine we were all shell-shocked (or SAM-shocked) but once again there was very little conversation over the interphone and no radio traffic other than the normal position reporting by the various cells. The navigators weren't even playing the Scrabble-like game of Spill and Spell as they usually did while they navigated us to Guam. And then the sun started coming up in the east and I was treated to one of the most beautiful sights I have ever seen. It was clear and from left to right and as far as I could see there were long puffy white contrails revealing the flight paths of the B-52 strike force on its way home. It rivaled any photo or film that I have seen of the large formations of B-17s during World War II. I still remember the uneventful approach and landing after about 15 hours in the air. We didn't know it at the time, but we had penetrated the strongest air defense network ever encountered in the history of aerial warfare.

That's the end of my story.

I still see those long puffy contrails in the sun over the Pacific when I hear the national anthem and I often wonder – did I really do that?

We flew to Hanoi two more exciting times and, in fact, we were Green 1 on the night of December 29th, the last B-52D formation over Hanoi during Linebacker II. They only shot two SAMs at us that night as they had run out.

Glasgow Crew E-21 (L to R): P - Capt Norman R. Vine, CP - Capt William A Ehmig, RN - Capt George W. Koch, N - Capt Robert Karwoski, EW - 1 Lt Douglas "Whip" Wilson, G - SSgt Robbie Robinson.

Pucker Time Out of Guam
Bill Ehmig

I reported to Glasgow AFB in December of 1964 after attending CCTS at Castle AFB and Survival Training at Stead AFB. What a shock for a kid from the South and his new bride from Miami! Glasgow was pretty remote and the weather was pretty severe, but the base was as good as SAC could make it. The hobby shops and other facilities were first rate and if you were a hunter or fisherman you were in your element. I was assigned to the 322nd Bomb Squadron commanded by Lt Col Wallace Yancy. The 91st Bomb wing (H) had recently been redesignated that from its previous designation of the 4141st Strategic Wing and was commanded by a first rate officer, Col George Pfeiffer.

I signed in on December 5th and the Squadron made life pretty easy for me during the holiday period: get settled in housing, go to Supply and draw my cold weather gear, to Personnel for turning in my records, Finance for the same, and all the things you need to do when reporting on a station for the first time. I went into a fairly relaxed square filling routine the second week. I flew once and took the obligatory qualifying tests. I remember the day I had to get signed off on weapons preflight. We went out to an alert bomber with the Wing Weapons officer, and the pilot and copilot of the airplane. I was to go into the bomb bay and preflight the weapons and give the proper challenge and response answers to the Weapons Officer. It was also a stringent rule that you didn't enter the bomb bay without pulling the pins that actuated the hydraulic struts that opened the doors; a stray current could close the fast acting doors on you and cause all manner of higher headquarters investigations. I tried to remove the pins, and when I failed, the AC of the alert airplane said, "Here, kid, I can do that".

The pins were frozen in place since it was 38 degrees below zero and the wind was blowing at about 20 knots. No one, including the crew chief, banging on the pins with a 2x4 could remove them, so the Weapons Officer thanked the crew for coming out, thanked the crew chief, and we left. "What now?" I wondered. He told me to come down to the Weapons Office to a very secure room, and we talked our way through the preflight. Satisfied that I knew the drill, he signed me off and one more square was filled.

Anyway, with Christmas over and New Year's Day gone as well, qualifying began in earnest. I flew with five different crews and finally took my checkride and was qualified by Maj Fred Mims on Standboard. I was put on a young crew, E-21. The crew was: Capt Norman R. Vine, Pilot, 1st Lt Bill Ehmig, Copilot, Capt George Koch, Radar Nav, 1st Lt Bob Karwoski, Nav, 1st Lt Doug (Whip) Wilson, EW Officer, and SSgt Robbie Robinson, Gunner.

For a while, in the exalted environment of crews and experience at Glasgow, we were viewed as real rookies and maybe escapees from "F Troop," but Norm ran a good crew and George dropped good bombs and so we were finally accepted as SAC professionals.

After about 10 months of training flights, Chrome Domes and seven-day Alert tours, our airplanes started going back to depot for Big Belly upgrades and we, along with all the other crews, got to go to Matagorda Island and drop real bombs. We even did some close formation flying, but there was no word about when we would be tasked to go to the Western Pacific. Then one day, Norm came by the house on CCRR in his 1505s. That was a surprise, even more so when he refused a beer and asked my wife if he and I could speak briefly in private. I think Kathy knew right then that we were going TDY, and in fact, we were. I was only allowed to tell her "we were going TDY to the Western Pacific, probably for 180 days". We were not supposed to say Guam although she came back from OWC bowling the next morning with as much or more info as Norm had given me. We had a week to get prepared since we, a junior crew, were not flying a B-52 but riding as pax on one of our tankers. We decided to close the house, hand it back to the Air Force and send her back to Miami to live with her parents. Anything was better than another unnecessary Glasgow winter.

The crew got to Guam after a far too short stop at Hickam, and settled in as well as possible. After a couple of over the shoulder flights with experienced IPs and IRNs, we seemed to settle in a Red 3 or Blue 3 and flew every two or three days. It slowly became comfortable, at least as a routine, but not creature comfortable.

After four months we were signed off as a lead crew and flew our first lead flight, as Bone 1, in a mass gas down South. Everything was going according to plan. We hit our tanker and took our prescribed download in one gulp. About an hour from coast in, Bob Karwoski came on interphone, "Pilot - Nav, we have a fire in the bomb bay". Bob's normally calm deep New Jersey voice was at least a half octave higher and the quiet of the flight deck changed from business as usual to intense interest. Norm asked Bob to unstrap and look through the bulkhead door window to see if he could determine the source of the fire. Bob came back to him that he couldn't see it but sparks and flames were coming from the left side of the bulkhead, out of his view, but right in from of the left forward main gear. Norm immediately ordered all non-essential electrical gear shut down as there were two electrical J boxes (junction boxes with significant current flowing through them) just inside the forward left wheel well. Pulling out the Dash-1 and Emergency Procedures sections we both started pulling the appropriate circuit breakers while George shut down the radar, Doug shut down most of his jammers and other equipment and Robbie reported his FCS shut down. All that was done pretty quickly. At the time it seemed to take forever. Looking back it was probably no more than three-five minutes before we were mainly flying on Mark 1 eyeballs and needle and ball. Norm was hand flying and asked Bob if the fire had subsided; after all, we had 84 Mk 82s in that bomb bay. "Pilot, as far as I can see the fire is out. I don't see the showers of sparks I did before," Bob replied.

Norm got on UHF and reported to the Airborne Commander, way up in Red 1, what our problem was and what our attempted corrective action had done. He further said that we were going to stay in formation and hold on to what we had until we could make sure that the fire didn't restart. The Airborne Commander thought that this was reasonable, but wanted us to get on HF and talk to 3rd AD Command Post. We did so and when Norm stated, "six souls on board" it all seemed very real. They cleared us to continue, with the potential for diverting to a South Vietnam (Ton Son Nhut) base if we had further emergencies.

We pressed on toward coast-in. Everything was going as expected and although it took hand flying, Norm had trimmed out the plane perfectly and it wasn't that difficult. With no more appearances of fire, smoke or arcing, Bob and George reported coast-in and approaching the PIP. Norm called Red 1 again and said we wanted to change to Bone 3 and drop on target using timing off of Bone 2. Again, the ABC was okay with this, but wanted up back on HF to clear it with 3rd AD. We got their clearance to proceed. We changed to Bone 3 climbing to fly trail above Bone 2. While Bob and George built and checked timing charts for a large number of distances, Bone 2's gunner gave us distance with his FCS. From IP inbound we were watching our distance from Bone 2. Norm was hand flying the airplane and I was flying the throttles to hold position at 1000 yards.

At the time we saw Bone 2 release, both Norm and I called "Hack" and timing was started on both watches downstairs. After the correct interval, George released the armed bombs and got onto the visual bomb sight to see if any appeared to fall out of the box. Since apparently none did, that he could see, we all breathed a sigh of relief. We had been told by 3rd AD that after the formation turned off target to coast out, we should divert to U-Tapao. In those days, there were no B-52s stationed at U-T. It was strictly tankers, some Australian Canberras, and Royal Thai Navy S-2Fs and P-3Cs. I had been jealous of the Tanker guys for flying from a base that was less distant than Guam and certainly more interesting. Finally I was going to see it.

Since darkness was coming, we were a bit concerned that we had to have some of the electrical power back on, enough for flaps for sure.

We both studied the approach plates for U-T and noticed the low hills to the north that were a bit worrisome - especially if we were essentially blind without radar. We started bringing on essential electrical systems slowly and gingerly and Bob watched through the porthole for any sign of fire or arcing. I kept the alternators isolated off the main bus and we were able to bring back all of the systems essential to approach and landing. We flew a straight-in from the sea and landed normally. We taxied in and were parked in an area that was clear of normal operations. The tanker OMS Line Chief told us when we got out of the plane that there was evidence of serious electrical arcing and fire in the junction box in the forward left wheel well. I think it was J-46, but the years have erased the certainty of that.

Norm called Guam and talked to Col Pfeiffer and Col Yancy and was told that we had done a fine job. We were taken to a couple of meat coolers (air conditioned trailers) and sacked out. The next morning, still in stinky green bags, we went to Supply to get some clean clothes. They issued us Vietnam style jungle fatigues, and for some reason, jungle (canvas) combat boots. A quick trip to the BX to buy skivvies, razors and toothbrushes made us more presentable, and the little Thai ladies with old fashioned sewing machines, had dressed our combat fatigues with embroidered names, wings and rank in a very short time.

There were essentially no B-52 parts or spares on U-Tapao at that time, or qualified B-52 maintenance guys to repair our airplane. We were pretty sure we'd get put on a tanker to return us to Guam, but 3rd AD told us to wait at U-T until the airplane was fixed and fly it back ourselves. It ended up being a four-day wait and life there was pretty good.

After getting back to Guam, I think all of the copilots I knew were jealous of my combat fatigues and my new, lightweight boots, which I was not allowed to fly in, as they were not SAC reg.

After reading about Crewdogs shot down in Linebackers I and II in the previous three volumes of *"We Were Crewdogs"* I realize that this incident pales in significance to those situations. At the time, it seemed a pretty high pucker factor waiting for the fire to torch off the bombs. We were all glad of the less sensational outcome.

Charles Haigh

*Best
To You,
George!
Charlie
"
12/08*

CINCSAC General John 'J. C.' Myer presents the author The Silver Star.

LB II: Birthday at Angels Three Zero
(From "Freedom Fighters," a larger work in progress…)
Charles Haigh

I completed a one-year in-country tour in June '69, flying AC-47 Spooky (Puff the Magic Dragon) gunships out of Bien Hoa AB, South Vietnam. Getting paid for killing commie bastards and flying almost every night in airplanes older than I was seemed kinda cool at the time. Even cooler, VC 'Charlie' had placed bounties on the heads of Spooky pilots. The assignment had led to a DFC and a chestfull of Air Medals, so - career-minded lieutenant that I was - I applied for a consecutive tour. My roommate, who dreamt only of flying front seat in an F-4, said it would be like back-to-back root canals without Novocain. He also believed there were no ugly women at 2 am and that Winstons tasted better than Marlboros, so what did he know?

I was assigned instead to SAC to pilot the mighty B-52H at Minot AFB, North Dakota. Unable to borrow a sharp knife from anyone - mine was dull from throwing at photos of Ho Chi Minh and Jane Fonda in the crew lounge, and I wanted a quick, painless death - I uttered a few choice words, threw a water buffalo t-bone on the barby, drank another six-pack of 33 (ba-mi-ba) Vietnamese beer, chain-smoked another pack of Marlboros (hard-pack only, for the flight suit pocket), and meekly accepted my fate. Meekly, if you consider throwing things and screaming at the top of your lungs for five minutes meek. It all counts for 20, I assured myself, pouring beer over my head and salt-encrusted flight suit. I had heard that 33-brand beer had formaldehyde in it, so maybe the beer would do for me what my inability to borrow a sharp knife couldn't.

On my way across the pond, I stopped on leave in Hong Kong to marry a lovely Chinese girl I had met on R&R. I thought she had money and she thought I was a rich American, so we were even. (We're still happily married after 38 years.)

After a BUFF training tour at Castle AFB, I reported to the infamous 'Northern Tier' base where, I had been warned: 'The wind doesn't blow at Minot—it sucks.' I also heard: 'There's a girl behind every tree,' and, when I arrived in January of 1970, in 40-below-zero chill-factor conditions, I saw a sign on a cement telephone pole that read: 'North Dakota State Tree.' Hmm...maybe there was a skinny Eskimo girl behind it? A bumper sticker on a car told me to 'Ski North Dakota!' Turtle Mountain, elevation 300 feet, was the highest point in the state. It was going to be a lonnnggg tour.

What the hay, I was blissfully married and, making the best of two days off the first year from flying, mission planning, simulators, commander's calls, SIOP briefings, TDYs, and back-to-back seven-day alerts, got my wife pregnant twice and had two sons.

Back in the armpit of the U.S. after an uneventful copilot Arc Light tour out of U-Tapao, I checked out as an AC and, lickety split, was on my way back to Castle to check out again in the B-52D for my second Arc Light tour. That time, I was headed for lovely Andersen AFB on Guam, the 'Pearl of the Pacific,' or, more commonly, the 'Rock.' I first heard the term, 'slicker than owl's shit' when someone described driving on Guam's roads that were made of coral, that somehow still had living matter in it (?) and so got 'slicker than...' when it rained, which was every 15 minutes. That's also when I first heard the term, while playing six-pack tennis: 'If you don't like the weather on Guam, wait 15 minutes.' Later, on our first takeoff from Andersen's undulating runway, I hoped my copilot wasn't lying when he assured me: 'Trust me, pilot. There's more runway once you get over that hill.'

Look, I don't remember all the exact times, dates, altitudes, etc. I wasn't anal-retentive (a-r) like my RN, bless his ever-lovin' heart, who kept stats on everything (17.4 officers for every stewardess on the island), so I didn't keep a logbook like I should have. I was too busy drinking all the beer on Guam, smoking way too many Marlboros, and doing things I don't even remember doing. I do remember it was a tour that was supposed to end in October of 1972, then November, then somehow was extended again to December, and then, all-of-a-sudden, it was time for a little pucker-factor adventure that would go down in history as Linebacker II.

My crew and I flew four missions over Hanoi and Haiphong during LB II and obviously survived all four, and all without so much as a single shrapnel hole in any of the four aircraft. (What did I do wrong?) We were fired at by beaucoup SA-2s, with too-many-to-count uplinks, got rocked around plenty—almost did a barrel-roll once—and I was temporarily blinded several times by near-hits, so maybe it was pure dumb luck on our part that we weren't blown out of the sky. To say it was exciting is an understatement. I remembered Winston Churchill had said during the Boer War that 'the greatest thrill in life is being shot at with no result,' and I knew where he was coming from. I even fantasized at times, as I had while flying Spookies through hails of tracer bullets from VC ground fire, that I wouldn't mind a 'small' hit by a tiny, relatively-painless bullet or piece of shrapnel, in an unimportant part of my body - a little toe, perhaps - so I could get a Purple Heart. My dad got a Purple Heart at Anzio, and I wanted one, too!

Anyhow, I like to think the reason we escaped unscathed was because I risked court-martial on all four LB II missions by taking Evasive Action As Required (EAAR) on every single bomb run, and for which I don't harbor an iota of guilt. I had survived 220+ night low-level close-air support missions as an AC-47 gunship pilot, fired over 2,000,000 rounds of 7.62 at commie bastards, been shot at hundreds of times by commie bastards without ever being hit, and I wasn't about to let some starry-eyed full-bull desk jockey in Omaha, Nebraska, tell me to fly 'straight and level' on the bomb run while SAM missiles are eating my ass (and my crews' asses) for lunch.

This ludicrous order reminded me of a Bill Cosby toss-of-the-coin routine called "The Brits and the Settlers," in which Cosby, playing a referee, says something like: "Okay, settlers, you won the coin toss. What will you do? Okay, you will hide behind rocks and trees and wear any color clothes you want. Brits, you will wear red and march in a straight line." This routine cracked me and my roommate up at UPT, but it wasn't funny now when I was being ordered, in effect, to "wear red and march in a straight line" while SA-2s are bearing down on me, my crew, and a B-52 loaded with bombs and jet fuel over one of the most hostile countries on earth.

This isn't false bravado; it's a simple statement of fact: if someone shoots at me, Maynard, in this or any other war, I WILL evade, and we'll talk about it later. Court-martial me, if you must. As the saying goes: "Better to be judged by twelve, than carried by six." Besides, we always got our bombs on target - BINGO! - because that's Job #1, and that's what professional aircrews do. Once more, in this order: We evade the damn missiles that will kill our asses (and prevent us from accomplishing our mission)—THEN and only then, we bomb the f-ing target.

The mission this little story is about is one we weren't meant to fly, and it's one for which I was awarded the Silver Star, personally pinned on by CINCSAC General John 'J. C.' Myer (see photo), and my crewmembers each got the DFC. It went something like this, allowing that I didn't keep a logbook and this all happened a long time ago. I do have an audiotape of this mission, courtesy of my wonderfully efficient RN, and I'll include a few choice remarks from that tape.

We were a spare for this mission, engines running and loaded for bear, in case someone crapped out of the lineup. Someone did - one of my UPT classmates at Moody AFB, in fact - and we dutifully filled the gap. I said dutifully, not eagerly. I had read somewhere that the definition of heroism was ordinary people doing extraordinary things under impossible conditions. We didn't feel like heroes at all. We just felt that, like a fireman entering a burning building to rescue someone, we were doing what needed to be done, when it needed to be done, and hoping like hell we'd live to do it another day. In other words, trite though it may sound, we were just doing our job.

Right after takeoff and gear up, "you've got it co," light the first of 80 Marlboros for the flight - Hello! - the number seven engine FIRE light comes on. I don't have a pilot's checklist with me as I write this 35 years later, but we did the right thing, cussed a purple streak, ran the emergency checklist, shut the mother down, and called the wave lead. "Dump your bombs in the ocean; burn off fuel and RTB," he told us, or words to that effect. "Sir, I've discussed this with my crew, and we want to press on with seven engines. We'll fly lower than everyone else to keep our speed up, but, with all due respect, we intend to press on." Silence, then something like, "It's your call. Let me know if, at any point, you can't hack it, and you can turn back." I'm thinking, "I didn't come this frigging far to turn back, and (with bravado I would come to regret) I have yet to meet anything I can't hack," but I said "Yes sir" and we pressed on.

The first aerial refueling begins and, suddenly, the possibility of not hacking it holds new meaning. One engine shut down, full bomb load, night, heavy weather, turbulence, spatial disorientation, profanity you never thought existed, over-temping ALL of the remaining engines, feet sloshing around in an inch of sweat in my combat boots, and, on my command, copilot on the controls with me. Somehow we managed to complete the refueling, drop off the tanker, and head for Hanoi.

(Note: what follows is partly remembrance from 35 years ago, partly from an aging, hard-to-understand audiotape of this portion of the mission, and partly from literary license for dramatic effect. Please forgive any glaring inaccuracies in what follows):

I remember we coast in from Laos, then ALL HELL BREAKS LOOSE. We're #3 at 30,000 feet on seven engines, the rest of the cell is at 35,000. Meaning, we lose mutual ECM coverage from our cell and stick out like a sore thumb on the bad guys' radar.

There is more radio chatter than I've ever heard, over the intercom, intercell calls, Red Crown and others on emergency channels, numerous uplink signals, AAA explosions, SA-2 launches from all quadrants, possible MiG sightings, Wild Weasels firing Shrike and ARM missiles toward radar sites on the ground, planes taking hits and going down, beepers and 'May Day!' calls over the emergency channel, brilliant explosions in the cloud cover (a SAM exploding harmlessly or a plane just hit?), darkened planes zipping past us and under us without warning - What the hell was that? - a near mid-air with a B-52 taking evasive action.

Tail gunner screaming: "SAMs approaching fast, pilot. Three of them! Five o'clock! Now seven o'clock! Break right, pilot! Break right! No, make that left, pilot! Break left!"

"Your left is my right, gunner! Which the f... is it?"

Gunner: "Your left, pilot! Your left, now! Now!"

Copilot: "I see one at 5 o'clock, another behind us! Break left! Hard left, pilot! Hard left!"

Near barrel-roll hard left, then blinding explosions above us and to our right rock the plane, but do no damage.

Copilot stating the obvious: "They missed us.

Gunner: "Damn near got us, pilot. One passed between the fuselage and right wing."

Now we're on the bomb run. Fear and anxiety ratchet up a notch, but we've a job to do. Uplinks and visual SAMs all the way, evasive action to avoid them (EAAR) then wings level the last few seconds, bombs away - ON TARGET - hard right post-target turn (Whose idea was that?), more uplinks, more visual SAMs, more EAAR, another near mid-air from a B-52 taking evasive action.

A few anxious moments until the cell reforms, then, finally, wings level and Balls To The Wall (BTTW) to feet wet and the hell outta this godforsaken place with seven over-temped engines. Audible sighs of relief when the Nav announces we're back over friendly territory.

Amazing how quickly things calm down. Serenity, now. Nav confirms our cell's heading toward the next aerial refueling. We settle in comfortably behind #1 and #2 for the long boring flight to the refueling area and then back to Guam.

I tell the copilot he's got it and light up a Marlboro, my 60th since takeoff. Time for soul-searching, inner reflection, "Why am I doing this?" on what just happened....

God, we coulda gotten killed there a dozen times. Was it worth it? Hell, yeah. You bet. F-in' commies! They got no right to take over South Vietnam by force. The commie bastards in Hanoi are supplied and supported by commie bastards in China and commie bastards in the Soviet Union, and South Vietnam can't defend itself against all those commie bastards without help. Commie bastards have murdered 100 million people since 1917. 100 million! Now they want to subjugate South Vietnam through force and terror and murder anyone who gets in their way. Somebody has to stand up to naked aggression by commie bastards against a free and independent country. It's not a civil war. It's not about reunification. It's not about the "corrupt" government of South Vietnam - it's about the PEOPLE of South Vietnam. The people of South Vietnam have just as much right to be free and independent as we do. They have the same fundamental human rights to freedom of speech, religion, press, assembly, protesting against the government, voting for the political party of their choice, and all the other rights of free people, and no one has the right to keep them from those rights. No one! It's called Universal Freedom, Maynard, and everyone has a right to it. That's what you're fighting for, my friend: Universal Freedom. Like it or not, that makes you a freedom fighter, and this hellacious mission you just went through is what freedom fighting is all about. Get used to it....

Back to reality: Crewmembers are filling out busywork paperwork, filling in squares to impress the desk jockeys back on the Rock.

"What's the date?" the copilot asks over the intercom.

"Tuesday," the EW tells him, laughing.

"I meant the date, asshole."

"19 December 72," the a-r RN chimes in.

I stop smoking, stare straight ahead. "What did you say, RN?"

"The date, pilot? 19 December 1972. We took off on 18 December and now it's 19 December."

My eyes flush with tears. I begin shaking all over, then sobbing, really sobbing. Shoulder-racking sobbing.

The copilot leans over, pats my shoulder. "You okay, pilot?"

I shake my head no.

"Are you sick?" he asks.

I shake my head no.

Wisely, he leaves me alone for a while. Finally, I recompose myself, light another Marlboro, take a deep drag. "I just remembered that December 19th is my wife's 24th birthday. I had forgotten all about it. What a terrible birthday present for a young Chinese girl from Hong Kong, still learning English, alone at Minot, North Dakota, at Christmas time, two kids aged two and six months, to hear that her husband, her sons' father, has been shot down and killed over North Vietnam on her birthday. What a terrible birthday present! What a terrible Christmas present!"

"But you weren't, were you?" the copilot replies. "You weren't shot down. You didn't die. You lived through it and will probably get a medal for flying a 15-hour combat mission with an engine shut down. I'd say that's about the best birthday and Christmas present ever."

Just as suddenly as I had burst into tears, I begin laughing. Not your ordinary laughing, but belly-ache laughing. Howling, out-of-breath, sidesplitting, foot-stomping, pee-in-your-pants laughing.

Too far from my suddenly-wise copilot to give him a bear-hug and the Wet Willy he deserves, I grab his left hand with my right and, laughing through my tears, squeeze his hand - hard.

Life is good, I remind myself. So very, very good.

The Rules Were Different at U-Tapao
Marv Howell

A Crewdog's General at U-T

Alex W. Talmant was a mustang, rising to Brigadier General through the ranks. In June of 1967 he assumed command of the 4258th Strategic Wing at U-Tapao Air Force Base, Thailand, and while assigned to the wing, he flew 100 combat missions over Vietnam. He was truly a Crewdog's General and went way beyond the call to make life at U-T more tolerable. When the 91st was deployed people began to hear stories about how well he looked after the crews. Some may be legend, but even so, he was great.

Transportation was limited so most of us either rode the base busses to various places or walked. A few had bicycles. The story was that General T told the staff if they were driving their cars and saw a Crewdog walking they had to stop and offer a ride to any base destination. I don't know anyone to whom that happened, but it made a nice thought.

One rumor was that an Airman on a bike ran in front of the General's car and was hit. Reportedly, General T jumped out and checked to see if the Airman was hurt. He was not and General Talmant grabbed the bike and threw it in the trunk of the car. He told the Airman to get in so they could get the hell out of there before the Air Police got there.

Another tale was of a Crewdog actually crawling home from the club. The General saw him and went to help. The flyer resisted help and both fell into the Klong between the club and the crew trailers. The guy telling the story said someone pushed him in. Later and sober, General T saw him at the club and related "the rest of the story." When he found out who rescued him, he made a public apology. It was worth a laugh.

When Son Krang (the water festival) was on, General T opened the briefing with an explanation that the custom of throwing water was a Thai way of wishing good luck through the coming year. He then asked if anyone had NOT had the pleasure of being doused. Most of us had, either from the hooch girls, the waitresses in the club or the bus drivers. One crew-dog raised his hand. Without a pause the General said, "After the briefing report to the back of the room." The luckless guy did as instructed to find a grinning Sergeant waiting with a big bucket of water.

One of the Stan board crew screwed up an approach or some minor thing, but it got reported. Talmant said, "Ok, you guys with the white hats keep wearing them so the guys can identify the hazard." He then said he had been on the phone all morning briefing 3rd Air Division and everyone else on the incident. He finished by turning to the Chaplain and saying, "Don't worry Chaplain, I also re-briefed God this morning too."

He joked with us, he drank with us, and he took our side. He truly was a Crewdog's General.

Proper Attire at the U-T O'Club

As everyone will remember, proper dress at the U-T club was shorts, shower clogs and some kind of shirt. Hats were optional.

After a flight one day, my crew and I and about 100 other guys were out on the patio, knocking back a few drinks and goofing off. I looked at my watch and it dawned on me that in every other Air Force O'Club they were announcing, "It is now 1645 hours. All those not properly attired must leave the club." "How ironic!" I thought and made a beeline into the club to find the PA system.

I announced "It is now 1645 hours. All those not properly attired will please leave the club." It was a big hit - loud laughs and applause when I got back to the patio. Then it turned serious. The Club Manager was asking everyone "Who made that announcement?" Someone said without thinking, "It was the guy in the yellow shirt." Now with that big a crowd you would think there would be many yellow shirts. On that day, there was only one and it was me.

The manager came over and started off with "You are going to be reported! You were in the room where the money exchanges are made and we now have to do a full inventory."

He asked my name and unit and I replied, "My name is Fred Thompson and I am from Columbus." He dutifully made notes and left.

Later General Talmant arrived. The Club Manager immediately grabbed him and in a very animated way obviously told him the story. He finished by pointing me out. The General knew our crew pretty well. He came over and sat down. He asked if I had made the announcement. "Yes Sir, I did."

I was thinking, "Oh sh*t, I am in deep trouble."

He then said, "Captain Howell, did you tell him your name was Fred Thompson and you were from Columbus?"

Thinking it was getting deeper, I said, "Yes General I did tell him that."

His stern face broke into a big grin, "Damn, that is the funniest thing I have heard this tour, what are you drinking?"

Just another example of how General T took care of the troops.

An International Evening at U-T

On my crew's first evening at U-T, we all decided to go out to eat. My copilot , Phil Mylenek, wanted to go to the Doll Court. The rest of the crew wanted to go someplace else. I decided to follow the two-officer policy and go with Phil.

We made it to the base gate and found a Baht Bus driver who knew where it was. I asked Phil what was so special about this place. He said, "Just wait and see. You will like it."

We were the only two in the bus and as we careened down the dark dirt roads I had thoughts that we were easy marks for a hit. I don't think Phil was at all concerned.

After what seemed like hours (actually about 20 minutes). We pulled into a courtyard and found a low building with a couple of signs. Again we were the only GIs in the place.

The 'special' part of the Doll Court was the menu. It was a Mexican restaurant! Not quite up to Tex-Mex standards but still good. We pigged-out on Tacos, enchiladas, and refried beans. We washed it down with several nit-noi Kirins and had a good time.

Thinking back it was truly an international event – members of an American aircrew, consisting of a Polish copilot and a Welsh EW, eating Mexican chow in a restaurant in Thailand.

To top it off, the same Baht bus driver was waiting for us when we came out.

Just one of those stories that made U-T unique!

The author, Kenneth Boone Sampson.

The B-52 Wheel Well Door
Kenneth Boone Sampson

It was April of 1967. We had launched from the island of Guam in the Pacific Ocean on a B-52 bombing mission to a target in Vietnam, Laos, or Cambodia. Guam has a very moist, wet climate and moisture gets into the bomb release shackles. Sometimes the moisture freezes into ice in the cold at our cruising altitude of 31,000 feet. This ice sometimes locks up the shackle so the bomb will not release at high altitude. But when the B-52 descends to a warmer, lower altitude on approach to our landing base, the ice melts, the shackle releases and the bomb falls on or through the bomb bay doors. It's a good way to blow up a B-52 and/or the runway on which we are trying to land. One time when we were rolling out from a landing, I peered through the bombing optics periscope and saw a bomb fall through the bomb bay doors and bounce on the runway from a B-52 landing behind me. The B-52 behind me ran over the bomb with the rear wheel trucks. Luckily the bomb didn't detonate.

We were the second B-52 bomber flight recovering into U-Tapao Air Base, Thailand, after hitting our target with 108 bombs. We had a warning light indication of a bomb hanger in the bomb bay, but knew that sometimes we would get a false warning light indication. It was the navigator's job to check the bomb bay in flight to be sure if we did, or did not, have hung up bombs. If there were, in fact, bombs hung up in the bomb bay we would try to drop them out over an emergency designated drop zone. We made a normal descent toward U-Tapao, leveled of at 10,000 feet altitude and depressurized. I left my ejection seat, departed the forward compartment, and entered the alternator deck with no helmet or parachute. The crawlway past the wheel well to the bomb bay is just 10 inches wide with fabric rope type handholds every six feet and the lights were very dim. Off interphone, I crawled along the crawlway to the bomb bay and observed that there were no hung up bombs. I then turned around and exited the bomb bay into the wheel well. In the wheel well I noticed an interphone station just adjacent to that big, huge right front tire.

I plugged my headset into the interphone station and called, "Pilot, Nav. There's no bombs in the bomb bay." The pilot thought that I was back in the forward compartment and so he put the landing gear lever down.

Just below me the landing gear door popped open with a loud bang and I found myself staring down at the blindingly bright, sunlit Gulf of Thailand! I was holding on to the old fabric rope grip with no parachute.

The next thing that happened was that the pilot realized that I was not in the forward compartment and popped the gear lever back up. The gear door accordingly slammed back shut with a bang and it was dark again.

I crawled back across the alternator deck to the forward compartment and reported to the pilot that I was back in my ejection seat. No more was ever said.

I checked the bomb bay in flight several times in later years but never plugged into the wheel well interphone again. I'll always remember that blaring, glaring, flash of light off the Gulf of Thailand.

A Seiko Watch, a Pair of Brass Candlesticks, and a Divorce
Tommy Towery

Since the days of Genghis Khan and Attila the Hun, warriors have returned from foreign lands laded with trophies of their great victorious battles. In the days of "rape, pillage, and burn" the pillage phase was one of the major tangible benefits of winning military campaigns and rewarded the warriors with collections of loot and treasures abundant in the distant lands. It was hard to bring back "rape and burn" souvenirs. Land warriors historically captured pieces of military weaponry, uniform parts, and small bits of pilfered art and jewelry items like watches and military medals. Many of the Crewdogs that flew B-52s in Southeast Asia grew up in homes which contained one or more war trophies their fathers had brought back from their own victories in Pacific or Europeans campaigns of World War II.

For members of the Strategic Air Command's bomber force, the ability to pillage through the remains of a battle was not available.

Since the bombers took off from a remote air base, dropped the bombs over a distant target, and then returned to the same remote airfield from which they departed, there were no spoils of war to be captured as war trophies. Thus, there were few alternatives left to fill the coffers of these gallant warriors who felt the need to return home with symbols of their victories – they had to buy them or get their buddies to buy them.

One of my first memories of Arc Light happened before I really even knew what Arc Light was. When I first arrived at Carswell AFB, Texas, I was hanging around the squadron building when a crew found out it was headed for a TDY to Southeast Asia. Almost immediately all of them had new best friends. Members of the crew found themselves surrounded by their fellow squadron mates who had small pieces of paper and checkbooks in hand. "Can I take your order please? Do you want fries with that Seiko?" At that time I seem to recall that the number one request was that the deploying airmen do the favor of buying their buddies Seiko watches. I can't remember the model number, but the favorite choice was the one with the moveable bezel with numbers that was designed for scuba diving. The going price was about $25 in early 1970 dollars. They may have cost less overseas, but it usually cost that to get someone to pick one up and return home with it for you.

And thus my introduction into my own world of SAC "War Trophies" began. I soon found out that the mysterious orient was full of all sorts of wonderful bargains, all awaiting the first G.I. with American Dollars in his pocket with which he was willing to part. And watches were just the tip of the preverbal good-deal iceberg. The next tale-tell sign of why money would soon be scarce was the first time I cracked the pages of the Army and Air Force Exchange Service (AAFES) catalog that was tailored for personnel stationed or on temporary duty overseas. The book was full of items at great prices but could only be ordered by military personnel stationed outside of the continental US. Being TDY qualified in that category. The AAFES catalog was a "wish book" for the wives of the Crewdogs, and a sure sign that any extra money that might come into the family budget through per diem from a TDY assignment would, in no short time, be returned to the military via the mail order catalog's offerings.

I learned new words that my Alabama upbringing did not teach me. There was this thing called a hibachi, which is common today, but back then was a new concept. The particular one of which I remember and speak was a big, green, egg-shaped ceramic one that no one would ever be able to afford to mail back to the states. But wait! Who had to mail things when we had our own airplane designed to haul things much bigger than a mammoth steak cooker! I learned on my first deployment overseas just how much stuff you could stick inside a "47 section" or into the many covey holes inside the crew compartment of a BUFF. The ultimate cargo container came to life when a plywood pallet was slung inside the bomb bay compartment. Many prayers were offered that an in-flight emergency did not require a redeploying bomber to have to open the bomb bay doors for any reason.

A steady stream of other items increased my vocabulary and depleted my savings. Rattan Papasan chairs, Mamasan chairs, wicker baskets and trunks, hanging basket chairs, and woven Princess chairs began pouring into the homes of the tenants of SAC base housing with the arrival of each returning crew. Soon they all looked like Pier One store showrooms, and most looked alike since they all shopped at the same place. No Crewdogs home was complete without a set of huge brass candlesticks that came from a catalog or a BX showroom or Officer's Wives' Gift Shop. While the ceramic elephant from Vietnam was primarily something that people actually assigned to Vietnam procured, some found creative ways to become proud owners of them, including me.

Guam is not a foreign country, since it is officially a US Territory. As such, items we found there on the economy were subject to taxes and probably import fees which made the price of goods as high as or higher than similar items in the continental states. Whatever the case, off-base shopping in Guam offered little to entice shoppers, save for the few small knick-knacks that some of the gift shops had. Guam's tourist economy was primarily geared toward the Japanese honeymooning couples who had to be able to fit any souvenirs of their visit to the tropical island into their suitcases for the flights back home. There were the traditional print shirts and dresses, wood carvings, and sea inspired jewelry such as the pink coral items. There were very few real bargains to be obtained on the Guamanian economy. The real bargains for the warrior shoppers on that Pacific island were found in a gift shop on base operated by the Officers' Wives' Club, the Andersen AFB Base Exchange, and any of the exchange systems on the Navy facilities there. For most it seemed that the Navy Exchanges held the greatest selections of loot.

When I arrived on that Pacific island in the early Seventies, stereo and camera equipment were the bargains of choice for me and most of my friends. One could wear a 35mm Nikon camera around the neck with the same pride as a "100 Mission" patch on a flight suit. The F-2 model Nikon was the ultimate camera for many, but Canon, Pentax, Yashica, and Ricoh were all more affordable. This was the time of interchangeable lens and add-ons galore for the serious photographer. Filters, tripods, cases, motorized film advancers – the list could go on indefinitely. There were also several models of range-finder cameras that were considered bargains, but the through-the-lens, 35mm, Single-Lens-Reflex (SLR) was reigning king.

The entertainment world had cycled into another era by the time I got to the Pacific. Like women's hem lines, stereo systems seemed to cycle between two alternate poles – one being the integrated system where all parts were contained in one box, and the other end being component systems, where each piece was a stand-alone part of the whole. At this point in history, the component system was the most impressive for anyone to own. One took pride in having a bookcase filled with many individual pieces of stereo equipment – all complementing each other to give the best sound available to the listeners. It is somewhat ironic that most of the people who bought all the high-end sound systems had long ago lost many frequency ranges of sounds to the shrill of high pitched jet engines constantly bombarding their ears. Wires hung like hanging gardens behind the shelves that held most of the component systems.

We were also going through several other skirmishes in the sound war. It was a battle between 8-track tapes and the smaller cassette tape systems. Quadraphonic sound was entering the market and large, wooden, Pioneer or Kenwood speakers fought for space with the smaller compact speaker systems. It was a time of reverberation components, stereo components, turntables, reel-to-reel tape decks, and frequency controlled colored flashing light boxes that predated disco balls. The exchanges were lined with row after row of different brands of these systems, all competing for the attention and ultimate purchase of a Crewdog. Each week it seemed that a new system with a higher model number and higher price tag would be added to the collection. If you didn't like the selection at Andersen, then you went the Navy exchange or to the one at the Naval Air Station.

Thank God it was also a time for neat and inexpensive earphones. It did not take long for one to get tired of listening to someone's 300 watt speakers in the next room or on the floor above or below you in the quarters that housed the crews. The base hobby shops allowed new tape deck owners to copy album after album of favorite musical tunes long before the discovery of Napster or other copyright violating systems appeared on the Internet. Such convenience of massive amounts of songs fueled the fires of many stereo system purchases.

While Guam was the Mecca of the stereo and camera world, U-Tapao and Thailand were not far behind in their appeal to the bargain hunters. The streets of Sattahip were lined with alternating rows of jewelry shops and tailor shops. Some of the streets were paved and some still were dirt roads. The jewelry shops were manned primarily with Thai merchants, but the tailor shops were owned and operated by Indian and Pakistani personnel and their entire families.

A custom tailored suit cost around $20 in the early Seventies in Thailand. That including being measured, a basic fitting, and then returning in two days to pick up the completed suit. You picked your material from a stack of bolts of various patterns, weaves, colors, and compositions. The tailor would stand you up and take your measurements which he wrote into a book. When you came back, he took his chalk and drew lines and stuck pins inside the basic framework of the garment and adjusted it to your exact desire. When you walked out of the door an hour later you knew that the next day you could take the Baht bus back to Sattahip and pick up the finished suit. I know this for sure, because I made a three day trip from Guam to Thailand during a break, flying on the Fleet Logistic Service (FLS) KC-135, and went to town to be fitted for a suit the day I landed and carried the new suit back with me on the flight that departed two days later. A whole book could be written on the Sattahip garment business when leisure suits became popular in the United States. Prices for a complete leisure suit, complete with flowery shirt, was about $18 – custom fitted. And during the heydays of tie-dyed t-shirts you couldn't miss finding a vendor near the gate.

No new suit would be complete without a new pair of shoes to go with it. The Indian tailors had relatives who were Indian shoemakers. I'll never forget walking into a shoe shop and having a salesman have me take off my shoes and socks while he opened up a green ledger book and placed it on the floor beside my bare foot. He had me place my foot onto a blank page and he drew the outline of my foot onto the paper with a pencil. He measured the arch of my foot and a few more things and he wrote the measurements into the book in Sanskrit. He wrote my name on the page, and then I told him what kind of shoe I wanted, picking the style from a stateside catalog and the leather from samples. The favorite leather at the time was elephant (or so they said) and boots were one of the most popular styles of shoes. I had a custom pair of elephant leather dress shoes done for me, which I picked up 24 hours later after I paid the going price of $10 to the merchant. Those shoes are still in my closet today, 37 years later. You can't wear out elephant leather. I'd bet that somewhere in that foreign country there is still a book with the outline of my feet on two of the pages.

It did not take the Thai merchants long to figure out how to attract the military spenders – girls and booze. Most of the shops had very cute Thai girls working the cash registers or showing the merchandise. And the door would barely be shut behind you when a Singha beer was thrust into your open palm. I personally would always get a Pepsi, but you could sit and drink beer and colas and shop as long as you wanted in the stores – all for free. Well, it was not really free, since most of the time you usually bought something while there. And the girls were friendly, but not selling themselves. At least the ones in the shopping district didn't.

Jewelry was the greatest bargain of the Thailand economy. Not only were there watches, but the jewelry stores were filled with case after case of rings, necklaces, and bracelets. If you wanted a black sapphire ring and you didn't see one that you liked, the jeweler would reach under the counter and bring up a cigar box full of white envelopes. He'd open the envelopes to show you hundreds of bare stones, and you picked out the one you wanted and told him how to set it. The next day, the ring was ready. If you paid over $25 for a gold and precious stone ring you were being suckered in. Yes, they did believe in the non-American practice of bargaining. My wife at the time collected turtles and I went into a shop one day and drew on a piece of paper a ring, necklace, and earrings shaped like turtles and made from opals and black sapphires. He took out the cigar boxes and I picked out the stones I wanted to use. It took them one day to make them, and the whole set cost me less than $50. A $30 emerald ring I bought her was appraised for $800 back in the states. Princess rings and multi-colored Jade items were also very popular, with Burmese Jade bringing the best price. If you wanted something pretty and cheap, Beggar's Beads were the solution.

One of my favorites in the jewelry stores was the Puzzle Ring. You could get four, six, or eight bands in white or yellow gold. The rings would fall apart when you took them off, but stay linked together as interconnected bands, and you had to know the secret way of getting them back together to be able to wear it as a ring again. Hence the name. The Urban Legend was that they were made to keep people from taking off their wedding rings when they were away from their spouses. It would work for many, and some got so frustrated that they had their's soldered together to keep it from ever coming apart. I wonder how many of these are still lying in drawers today awaiting someone who knows the secret to getting the ring back into one piece.

Gold "Four-Seasons" bracelets were the ultimate big purchase items for many foreign shoppers. They were intricately carved, heavy bracelets with lots of 14k gold. I recently saw one at auction that was described as having 73.5 grams of gold and being purchased overseas in 1966 and had an asking price today of $3,895. The prices in Thailand for the bracelets varied over the years I visited there, primary because of the fluctuating prices of the gold market, but in 1972 I could have bought one for $220. That was a lot of money back then, so I bought a silver one instead for $25. The other gold bargain was Baht chains. Measured as the weight of a baht coin, you could buy one, two, or three baht necklaces, and many smart shoppers jumped at the opportunity of picking up those items.

Carved teak wood serving bowls were cheap souvenirs for the women in our lives. Mothers especially liked them. Salad bowl sets and odd shaped bowls like fish designs were common. Another shiny object that caught the eyes of many-a-shopper was the Thai Bronze dinner flatware sets which also sometimes had teak wooden handles. None of the male shoppers ever considered the amount of elbow grease it would take to keep those things cleaned and shinning, and I don't think I ever ate with any of them at any of my friends houses – even the ones that I knew bought a set.

Another oddity of the shopping world of Thailand was art. Yes, there was culture there too. There were many places you could buy oil paintings at very reasonable prices. The one piece I wish that I had purchased, but did not, was a wonderful painting of Moses and the Ten

Commandment tablets. The reason I wanted it was that it was not just an artist's idea of what Moses looked like. It was the real Moses. It was Mr. Charlton Heston himself and a scene right out of the movie *"The Ten Commandments."*

There was a booth at the Officer's Club where you could get custom paintings done from your own photo. I had a photo of the Piper Tri-Pacer I owned turned into an original oil painting for about $20. There were many other spoils of war for the true art connoisseur. One of the oddest and funniest things that happened in the art world was when deployed Crewdogs or staff pukes would have paintings done of their wives or girlfriends made from pictures they carried in their wallets. Many times I walked past the artist manning the booth to see a photo of someone's wife or girlfriend paper clipped to a centerfold out of *Playboy*, and watched as the artist took the body of the Playmate and adorned it with the face of the young lady in the picture. Sometimes it was comical but sometimes it turned out really professional. That was the days before Photoshop and other digital imagining software. I don't know how the individual felt about the whole world seeing his wife or girlfriend completely nude and on public display at the Officer's Club while awaiting him to pick it up, but sometimes the paintings stayed on display for weeks. I also don't know how the wife or girlfriend back in the states felt about the body change she was given in the process! I often wonder if any of those paintings still exist.

The running joke during my time was that the three things you got on an Arc Light trip was: a Seiko watch, a pair of brass candlesticks, and a divorce. That was true for way too many whose marriages could not hold up to the demands of the service and the excessive trips TDY.

No matter how long the TDY was, or what the mode of transportation to and from the overseas bases, the shopping bug hit just about everyone in one way or another. Every bird deployed back to the states was filled with boxes of treasures. Other periods and other wars and other TDY locations had their own stories and loot, but I'd bet almost every Crewdogs still has something in his house that came from an overseas TDY trip.

I still laugh at the copilot I had on my last trip to U-Tapao. His wife wrote and wanted him to buy her a watch since all the other guys on the crew had already sent their wives watches. Rather than going downtown to Sattahip to get her a Seiko or Citizen or Casio ladies watch at a great price, he went to the Base Exchange and bought her a "Made-in-America" Timex that "Takes a licking and keeps on ticking." He could have gotten her the same one at the same price in the BX at Carswell!

There were some folks that just should have just stuck to the "rape and burn" scenarios and left the pillaging to us pros.

Engine-Running Spare
Gery Putnam

It was a dark, wet, and hot night at U-Tapao, Thailand in the Spring of 1969. We were the engine-running spare on the apron at the end of the runway opposite from where the cells would launch. For some reason I had a substitute copilot who was the son of an Air Force general as I recall. The first cell launched and after the third bird had roared down the runway for what I considered a sufficient amount of time to be airborne (the rain precluded visual observation), I vented the cool air to the lower deck and opened my side window. Within a few seconds we heard Charlie on the radio query "What happened (whatever color) three?" I don't remember the response, but knew we would be launched and attempted to close the window. As Charlie directed us to launch, I instructed the copilot to taxi onto the runway while I used my foot to coax the window shut. I then took the controls and proceeded to take off.

Immediately after becoming airborne, I shifted to the instruments and found myself with needle, ball, and airspeed with lots of warning flags displayed. Pitch black, no horizon, no boat lights, no stars, the pucker factor went way up. My first reaction was thinking a circuit breaker had popped and suggested that the EWO come forward and check the panel behind me. As I remember, his response was "You want me to get out of my seat?" Naturally we notified Charlie of our predicament and received the standard "standby" response. But then he added "Try recycling the master switch." Duh, that should have been obvious. When I reached down to recycle it, I discovered it was off. Turning it back on did wonders for our instrument panels. Apparently when I used my boot to close the window, I kicked it off as I brought my foot back down. I told Charlie "Thanks, I owe you one" and proceeded with the mission.

That, by itself, would be unremarkable except for a postscript. A few weeks later that same substitute copilot, flying with his assigned crew, took off from Guam, turned right instead of left and plunged into 6,000 feet of water. No survivors. The guesstimated cause was loss of instruments. I found it difficult to believe that the copilot would not have detected the fact that they lost instruments having recently experienced the event. I did express my belief to someone in authority, but don't know if it was ever considered in the investigation.

Chapter Four

Lest [lest] - *conjunction* - for fear that.

We [wee] - *pronoun* - oneself and another or others.

Forget [for-get] - *verb* - to cease or fail to remember; be unable to recall.

Long Live B-52 Gunners
Richard L. Gaines

This story is not about me, but about my gunner, Master Sergeant Gene Gries, SAC's best. First, let me say a word about gunners. They were the only enlisted members of a six man B-52 crew. In the older models their flight position was in the extreme tail end of the fuselage and they operated their quad-50 caliber machine guns directly. In the B-52G and B-52H the gunners were up front with the other five crew members. In these models, they became a closer member of the crew. As far as we were concerned they had the status of the other officers. Their job was to work closely with the Electronic Warfare Officer in defense of the bomber. B-52 gunners shot down two North Vietnamese MiGs during the war, without the loss of a single bomber to enemy aircraft. Most MiGs were wary of attacking, opting rather to fly at the bomber altitude and heading. This information was radioed to the North Vietnamese SAM (surface-to-air) missile controllers to improve their SAM effectiveness.

Back to Master Sergeant Gene Gries. Gene was my gunner for almost three years at Grand Forks AFB, North Dakota. He was very proficient, reliable and a good guy, so I always looked forward to taking him along with me when our time "in the barrel" to Vietnam came. Actually, the bombers flew out of Guam, Okinawa, and U-Tapao AB, Thailand, flying four to 12 hour bombing missions into Vietnam.

Gene was transferred to Beale AFB, California, in early 1969. His second assignment to U-Tapao was with a Beale crew during 1972. On the fateful day, his crew was on a mission out of U-Tapao with two other B-52's. Over the Thai/Cambodia border, their airplane flew into a very severe thunderstorm. Within seconds the B-52 was in serious trouble. Violent turbulence caused the crew to lose control of the airplane. They were almost inverted when Gene saw the red BAILOUT light illuminate. He then saw several crewmembers, still in ejection seats, pass his tail gunner position. Little did he know that a collision of the pilots and copilots seats had sealed their fates. The other three crewmembers failed to eject for some reason. Gene jettisoned the entire tail gun pod with no trouble, but because of the steep dive, he was unable to get out of the large hole. Think about climbing out of an automobile sun roof as the car flips end over end, only much more difficult. He tried over and over to pull himself up and out of the doomed bomber, but could not.

He sat back in his seat with some difficulty and thought to himself (his words): "Sylvia (his wife) is really going to be ticked off if I don't get out of this thing". With one extreme effort, he pulled himself, his parachute and a 60-pound survival kit up and out of the hole. The parachute opened as advertised at 14,000 feet and Gene released the survival kit lever to let it fall 30 or so feet below him. That was very difficult because he was still in the severe storm that had caused the B-52 to break up in the first place. Finally, he broke free of the clouds enough to see lights on the ground and just as quickly his parachute rose back into the raging storm. Pelted by hail, heavy rain and severe turbulence, he did that over and over for an estimated 30 minutes before finally coming to rest in a rice paddy.

The fun was just beginning. Several farmers or soldiers came toward him, illuminated by the constant lightning. They seemed to hesitate every time he talked on his survival radio to the rescue helicopter, which arrived before the locals got to him. By that time the weather was really getting dicey. The helicopter dropped their sling to him and just as he secured it to himself, severe wind hit again, causing the helicopter to become uncontrollable. Gene was dragged hundreds of feet across the rice paddy and almost drowned. The helicopter recovered and winched the cable up, getting him aboard. Gene was taken back to U-Tapao, Thailand, without serious injury and then flown back to Beale AFB, California.

Gene visited me and my wife Bunny a few weeks later at our home in Merced, California. His hair had turned almost white. He flew one last time in a B-52 shortly afterward then retired from the Air Force. We visited Sylvia and Gene at their home in Las Vegas in 1993 and had a great time telling war stories. Sylvia died in 1997 but we still keep in touch with SAC's best gunner. After all, Bunny and I were made "Honorary Gunners" years ago.

Note: SAC did away with the B-52 gunner position in 1993 and removed all guns from the remaining airplanes, relying on the wizardry of the Electronic Warfare Officer to protect the bombers. SAC itself was disbanded a few years ago but even with a greatly changed mission, the great B-52's fly on. May they do so forever with the ghost of SAC's gunners flying in that empty seat.

Stratofortress
George Thatcher ©1998

A crewdog must be extra tough,
To strap his ass into a BUF,
One ride, for most, would scratch their itch,
To fly that ugly son of a bitch.

The T.G. meter zeroes,
Bombs away, that eerie sound,
And thirty tons of iron death
Go tumbling toward the ground.

Who knows how many perish,
Or if we've merely ground
Miles of jungle into toothpicks?
Never time to hang around.

It's eerily impersonal,
We're mere technicians: flunkies!
Like, instead of killing men, we're
Making orphans out of monkeys.

Or, if you buy the bullshit
That comes down from levels highest,
We're the noblest of heroes,
Killing Communists for Chri-est.

Starboard turn, make for the coast,
Six hours to Guam; you're dying
in your seat of utter boredom.
And we call this "combat flying."

The Combat Crews
George Thatcher

They prowl the skies, a bold, contentious lot,
Diverse their nature, ethnic polyglot
of color, culture, race, and every creed.
Their commonality: the red they bleed
in line of duty, for the higher cause
for which they take the oath, with ne'er a pause.

No strangers to the dark side of mankind,
In matters of the heart, quite often blind.
Their duty's needs pre-empting tenderness,
After all, how else do real men deal
with stress?

But underneath the silver of those wings
that they wear, with their oft-concealed pride,
Beat hearts just like so many that have died
in the line, for some vague but noble goal,
Like keeping children safe and families whole.

Though we admit too seldom, often late,
The sacrifices made, but rarely seen,
A safer world will yet appreciate
the combat crews, who fly in olive green.

The Boneyard
George Thatcher ©2001

We stand rigidly
on our marks,
At attention
in aeternum,
Lined up
in endless rows,
Tip to tip,
Staring fixedly
across the miles
of dusty drab.
Waiting.

We are
at the ready
still,
Although
some of us
now say the call
will never come.
After so many years,
So much hope,

We are beginning
to quietly
crumble
from inside.
But we
hold ourselves
Erect.
Alert.
Because you
never know.

Oh,
sometimes they come,
To inspect,
To reprise,
Or just to ponder.
Eloquent in their
wordless
concentration.
Do they remember?
Are those
the tears they
couldn't shed
before?

They
are older now,
But they
are ready too.
Ready for the call.
And
they honor us
by remembering
that we served
together,
Man and Machine,
As one.
A team.

You there!
Try me out!
Climb aboard and

make me live
again.
Make me roar,
surge,
and flex.
Trust me!
For I never
failed you then.
And
I never will.

Free me!
Fly me!
Just once more.
Just this once.

Please.

Semper Fly
George Thatcher

I'm often reminded,
as life hurries by,
that I've never quite lost
my desire to fly.

Walking through airports,
strapped in a seat,
takeoffs and landings,
still hard to beat.

Smell of the jet fuel
floats in the air.
Roar of the engines,
destined for where?

War stories bubbling
from deep, hidden springs,
and they say, "There I was…"
"Comin' in on a wing…"

Thoughts of old crewmates,
most of them gone.
Names live forever,
they're waiting beyond.

And have I been spared
just to carry the tale?
Searching for answers
to little avail.

But I dream there's a Plan
for us all, that tells why.
Maybe under my name
it'll say
Semper Fly.

Résumé [rez-oo-mey]
- *noun* - a brief written account of personal, educational, and professional qualifications and experience.

The Contributing Authors

Ben Barnard

Colonel USAF, (Ret.) During my Air Force career, my flying assignments took me to Carswell, Blytheville, and Griffiss. Staff assignments included the F-16 System Program Office at Wright-Patterson, the DO's staff at SAC Headquarters, AFROTC duty at Miami University in Ohio, Attaché duty in Mexico City, and Defense Intelligence Agency Headquarters in Washington, D.C. I accumulated over 4,000 flying hours in various models of the B-52 and served as instructor pilot, flight commander, operations officer, and commander of a bomb squadron and commander of an operations support squadron.

After completing 30 years on active duty, my wife of 37 years and I settled in the Hampton Roads area of Virginia. I am more or less still in the military as my post retirement job is senior military analyst with the United States Joint Forces Command.

My wife and I have three adult children and five grandchildren with number five on the way.

James E. Bradley

Lt Colonel, USAF (Ret.) James E. Bradley graduated from college in 1959, with a Bachelor of Science in Mathematics and Chemistry, and enlisted in the United States Air Force in order to avoid being drafted and to preserve a chance of getting into flying training. He entered basic training in July 1959 at Lackland AFB, San Antonio, TX. He completed basic training in October 1959 and was assigned to the Air Force Special Weapons Center, Physics Division's Pulsed Power Laboratory at Kirtland AFB, NM, arriving in October 1959. There he performed Engineering & Scientific Aide duties consisting of mathematics and computer data reduction aide. This was his first introduction to computers, key punch machines and to scientific laboratory work. The Executive Officer in the Physics Division was Major Lew Allen, later he became General Lew Allen, Chief of Staff United States Air Force. The focus of the work here was finding a way to simulate high-altitude nuclear detonations in a laboratory. This was desirable because the banning of atmospheric testing of nuclear weapons was eminent. He was promoted to Airman 2nd Class during this assignment.

He applied for and was accepted to Officer Training School (OTS) and entered training in April 1961 at Lackland AFB, TX, and after three months of training was commissioned a 2nd Lieutenant on 27 June 1961. From OTS he was assigned to Undergraduate Navigator Training at James Connelly AFB, Waco, TX. Here he flew on the venerable T-29 Flying Classroom. He received his Navigator's wings in April 1962.

He was assigned to Electronic Warfare Officer Training at Mather AFB, Sacramento, CA, and graduated from EWO schooling in February 1963. He attended USAF Artic Survival Training at Stead AFB, Reno, NV, and entered B-52 Combat Crew Training School at Castle AFB, Merced, CA, in March 1963. All flights were on B-52B/F aircraft. Completed training in July 1963 and was assigned to Glasgow AFB, MT, where the B-52D aircraft were assigned.

He remained assigned to the 91st Bomb Wing, 322nd Bomb Sqdn as an EWO on a Combat Crew until the unit was disbanded in spring of 1968. Lt Col Bradley has approximately 3,400 hours total flying time with 2,356 hours in the B-52B/C/D/F.

Lt Col Bradley was assigned to the USAF Iron Hand mission known as the Wild Weasel program. He flew in the F-105F as an EWO or Bear assigned to the 355th Tactical Fighter Wing, 333rd Tactical Fighter Sqdn stationed at Takhli RTAFB, Thailand. After the Wild Weasel tour he entered the Air Force Institute of Technology's Civilian

Institutions Graduate School at Texas A&M University to study Computer Science and graduated in December 1970. He was assigned to Headquarters Air Training Command in the Data Automation Computer Division where he was Computer Operations Branch Chief.

He attended the Armed Forces Staff College and graduated in June 1974. He was then assigned to the U. S. Readiness Command/J5, Joint Operations Planning System Officer, located at MacDill AFB, Tampa, FL, were he was instrumental in the development of the JOPS computer based planning system. From there he was assigned to Headquarters, Tactical Air Command (TAC) on the Planning Staff of the DCS/Plans, BG Larry Welch, later Commander-In-Chief SAC and subsequently Chief of Staff USAF. It is a small world.

After retirement from the USAF in February 1980, he went to work for the Boeing Military Airplane Company, Wichita, KS, on the B-52 Modernization Program. Here he was assigned as a Software Test Engineer in the laboratory where he was responsible for testing the Offensive Avionics System's software in the laboratory prior to flight in the B-52G/H. He also worked on flight simulators and mission planning projects as software test engineer. He retired in 1995. He and his wife, Marian, currently live near Westmoreland, KS, on the farm she was born and raised on.

John R. Cate

MSgt, USAF (Ret.) Was born and raised in Oliver Springs, Tennessee. Attended Clinton Senior High School, graduating in 1971. Attended Tennessee Technological University from 1971 until 1973. With the draft a certainty in my future, I enlisted in the Air Force in April of 1973.

My initial assignment was to RAF Bentwaters, England as an F-4D Phantom II Weapons Release Mechanic. Next assignment was to Lowry AFB, Colorado as an instructor with ATC. After serving as a classroom instructor for five years, I applied for the B-52 gunnery career field. My father had served in the Army Air Corp during World War II as a B-24 Top Turret Gunner/Flight Engineer with the 5th AF and I had always wanted to follow in his footsteps. After graduating from CCTS at Castle AFB, California, I was assigned to the 28th BMS at Robins AFB, Georgia, as a B-52G Gunner. In 1983 I was assigned to the 1st ACCS as an Airborne Radio Operator serving on the E-4B

NEACP. I was than allowed to return to the B-52 in 1986 being assigned to the 23rd BMS, Minot AFB, North Dakota as a B-52H Gunner. While at Minot AFB, I attended CFIC at Carswell AFB, Texas in 1987. I served as a B-52H Instructor Gunner and was assigned as S-02 Gunner. I was next assigned to the 43rd BW, Andersen AFB, Guam as the Wing Gunner. I had the honor of serving as the last 43rd BW Wing Gunner. After the deactivation of the 60th BMS of the 43rd BW in 1990, I was assigned as a B-52H Instructor Gunner with the 325th BS, Fairchild AFB, Washington. With the removal of the gunner position from all B-52s in October of 1991, I was assigned to the 964th AWACS, Tinker AFB, Oklahoma as an Airborne Radio Operator, flying the E-3B AWACS aircraft. I served in this position until 1993 when I was medically grounded and removed from flying status. During the last two years of my Air Force career, I served as NCOIC, 552nd ACW Wing Safety Office.

Retired in 1995, I returned to Clinton, Tennessee, where I am employed as an estimator for an established building supply. I have one son who is an architect, living in Dallas, Texas. Serving as a B-52 Gunner was the highlight of my Air Force career and it has given me a lifetime of memories. *C'est La Vie.*

Doug Cooper

Lt Colonel, USAF (Ret.), got his commission through OTS in 1965 after graduating from the University of U-T ah and facing an imminent first draft pick from the U.S. Army. He completed Undergraduate Navigator Training and Electronic Warfare Training at Mather in 1966 and 1967.

His first assignment was to Beale AFB, and the 744th Bomb Squadron. His crew, E-30, was one of the first selected to attend RTU in D Models at Castle in the summer of 1968 after which he spent six months between the Rock, U-Tapao and Kadena.

After leaving, Beale, he went to Carswell just in time for Bullet Shot and served an additional four TDYs to Guam and Thailand.

After an assignment to the Carswell Command Post (he was the first non-pilot to become a command post controller in SAC), he was reassigned to the SAC Airborne Command Post (Looking Glass).

Subsequent assignments at Zaragosa, Spain; Incirlik, Turkey; and Mather AFB were followed by retirement in Sacramento.

Doug and his wife of 39 years, Susan, live in Lincoln, California. His hemorrhoids still ache when he thinks about Giant Lance sorties, o

Bill Dettmer

Lt. Colonel, USAF (Ret.) Bill Dettmer was commissioned through Air Force ROTC at Rutgers University in 1966. He completed Undergraduate Navigator Training at Mather AFB in 1966-67, and Nav-Bomb Training in 1967-68. His initial operational assignment was as a B-52G navigator at Beale AFB, California.

In 1971 Bill completed Undergraduate Pilot Training and returned as a copilot to the 744th Bomb Squadron at Beale AFB, where he had previously been a navigator. He finished his second B-52 tour as an aircraft/crew commander in 1976 and went into the rated supplement as a Minuteman III ICBM crew commander, flight commander, and subsequently operations officer. In 1980, Bill returned to B-52s as an aircraft commander in the 2nd Bomb Squadron at March AFB, and subsequently Chief of the 22nd Bomb Wing Training Division.

When the 22nd Bomb Wing was converted to an air refueling wing in 1982, Bill was reassigned to Headquarters Air Force Logistics Command at Wright-Patterson AFB, Ohio, directing AFLC's participation in Joint Chiefs of Staff exercises. In 1984, he was assigned to Headquarters 13th Air Force (Clark Air Base, Philippines), where he served as chief of logistics plans and deputy chief of staff for logistics for two years, and chief of military civic actions for one year.

After reassignment to Headquarters 15th Air Force (March AFB) for a year, Bill retired in December 1988. He subsequently taught graduate courses in systems management for the University of Southern California for seven years. In 1995, Bill founded his own management consulting company, Goal Systems International, which he still heads today. He is the author of six books on various aspects of business management and strategy development.

Derek H. "Detch" Detjen

Major, USAF (Ret.) Was born in New York City. Grew up there and in the Northern Kentucky/Cincinnati area, entering the Air Force Aviation Cadet program at Harlingen AFB, Texas in late 1960. He achieved the rank of Cadet Major and was the Exec. Officer of Adams Squadron, receiving his 2nd Lieutenant bars and navigator wings in September of 1961. After finishing Electronic Warfare Officer's school at Keesler AFB, Mississippi where he met his wife Betty, it was off to survival training at Stead AFB, Nevada and then B-52 crew training at Castle AFB, California, before his first operational assignment to the D-model at Turner AFB, Georgia. In March of 1966, he participated in the first six-month Arc Light tour to Guam, flying in the first-ever "North" mission on 12 April 1966 with Ellsworth crew E-20, commanded by Lt Col Paul Corn. Returning to Guam in both 1967 and 1968 as a crewmember and staff EWO briefer from Columbus AFB, Mississippi, he was part of Crew E-13, led by Maj Joe Steele, becoming the first crew to complete 100 Arc Light missions in December of 1967. Crew E-13's 5th Air Medals were presented to them in the briefing room at Guam by the then CINCSAC, General Paul McConnell.

Assigned to Castle AFB, California's Replacement Training Unit (RTU) in July of 1969, he spent probably his most satisfying and rewarding four-year tour, training the G and H-model crews on Arc Light tactics/procedures before their certification and deployment to Guam in the D-model. Four years in the 1st Combat Eval. Group followed at Wilder, Idaho and Ashland, ME, including a five-month stint on Guam at Detachment 24 "Milky," training the recently deployed B-52 wing in ground-directed bombing procedures. His last five years were spent at Barksdale AFB, Louisiana, where he was in charge of B-52 and KC-135 EWO study. A notable event while there was watching the 1980 U.S. vs. Russia Olympic hockey game with the SAC Inspector General team, delaying the onset of their annual visit! He was selected for a final, memorable trip to RAF Fairford, England as part of the 1982 Crested Eagle NATO exercise. His military decorations include a Distinguished Flying Cross, eight Air Medals, a Meritorious Service Medal, two Combat Readiness Medals, an Air Force Commendation Medal and several lesser awards.

After his Air Force days, Major Detjen worked for General Dynamics on the Trident submarine at NSB Kings Bay, Georgia for five years, also attained a graduate degree from Valdosta State University and taught in their on-base education program. A final nine-year stint at Aiken Technical College in Aiken, South Carolina saw him coordinating both the Management and Marketing majors before his retirement in 2000. A lifelong devotee of The Masters golf tournament, he now resides in Evans, Georgia, about 10 minutes from the Tournament, which he has managed to attend for over 40 years.

Lothar Deil

Major, USAF (Ret.) was commissioned upon graduation from the Air Force Academy in 1976 with a B.S. in Astronautical Engineering. After UPT at Vance AFB, Enid OK was assigned to B-52Gs at Fairchild AFB. A medical condition ended his flying career in 1982. SAC decided he needed to work for a living and sent him to SAC/IN to build computer models for use in mission planning and threat assessment. In 1987 he moved all the way to SAC/DO as program manager for the Electronic Warfare systems for SR-71 and U-2, eventually becoming the technical lead for all SAC EW reprogramming.

He retired in 1993 and worked in information technology for the private sector, becoming the Chief Information Officer for a medical research firm in Seattle, WA. and then the Northwest Region Information Technology Director for a national engineering consulting firm.

In 2001, following 9/11, he returned to the Air Force to continue his service as a civilian electronic engineer with the 453 Electronic Warfare Squadron, Lackland AFB, TX. He is currently the Lead Development Engineer responsible for the maintenance and improvements for a variety of Air Force and Joint operational software including the Improved Many on Many Radar and Communications predictive software systems. He was just awarded the Air Force Engineer of the Year 2007 in the Senior Division.

Bill Ehmig

Former Captain, USAF (Honorably Discharged) Bill Ehmig () was commissioned in 1962 from OTS at Medina Base, San Antonio TX. He was stationed at Homestead AFB, FL as a Personnel Officer and submitted 8 applications for pilot training until he finally, after a lot of painful dental work at the Homestead Dental Clinic and a waiver from USAF, he was assigned to Class 64 H at Moody AFB for UPT. He and his new wife Kathy lived in a trailer in a cornfield near Ray City, GA for 55 weeks and then took TDY enroute to Castle and Stead before reporting to Glasgow AFB, MT. While a crewdog at Glasgow Bill flew an Arc Light Tour and a Port Bow Tour before Glasgow was closed down. After having orders for Beale and Seymour Johnson cancelled, Bill and Kathy went off to Castle for G&H difference school with a nine day old daughter (one of the last children delivered at the Glasgow AFB Hospital). They then drove with a young baby, two cats and a '65 Mustang from Merced to Warner Robbins GA, where Bill joined the 19th BW (H), 28th BS. Bill participated in another Arc Light tour while at Robbins. Upon resigning his Regular Commission in Sept 1970, he entered the civilian sales arena with a Johnson and Johnson firm, was promoted to Division Manager and recruited to US Surgical as a Regional Sales Director. Another recruiter beckoned after 4 years and Bill joined Amersham Corporation as National Sales Manager. In his 25 years at Amersham, Bill was General Manager Amersham Canada, Director of Marketing, Vice President of Pharmacy Operations and retired as Vice President, Professional and Industrial Relations. He served as chair on numerous committees of professional organizations including the Society of Nuclear Medicine, American College of Nuclear Physicians, the Council on Radiopharmaceuticals and Radionuclides, and others. He and Kathy spend their retirement in Dataw Island SC and Seven Devils NC. Bill's favorite activity is working at the Mighty Eighth AF Heritage Museum in Pooler, GA as a docent at the Mission Experience.

Dale Fink

Former Captain, USAF, (Honorably Discharged) graduated AFROTC at Texas Tech University with a B.B.A. in Accounting. Attended "Pre-Nav" training at Valdosta, Georgia before the "Great Washout of '73" and then attended UNT, Bomb Nav Training, and CCTS. Served 6 years as a B-52D navigator at Carswell AFB, Texas. After leaving the Air force, worked in several different industries (Banking, Marketing Research, Brokerage, Truck Manufacturing, Clothing Manufacturing, Nuclear Power Plant Inventory and Medical Technology) as a computer consultant/manager/ programmer for the past 28 years. Currently working as Program Manager and Chief Programming Architect for a company that provides online

Boating/Hunting/ATV certification in over 40 states. He and his wife, Lynn, reside in Arlington, Texas.

William D. Fritz

Lt. Colonel USAF (Ret.) Bill Fritz was a WW II army brat. After the war, he grew up in New Jersey and graduated from Rutgers University with majors in Chemistry, Physics, History and ROTC, and was commissioned on graduation in 1963. The first PCS, after UNT and Electronic Warfare Training, was the 397BW/596BS at Dow AFB in 1966, closely followed by the move by the wing on closing of Dow to Barksdale in the 2BW. Almost on arrival, his crew at the time (E-55) was assigned to an Arc Light tour from Feb 69-Aug 69. On return from Arc Light he was reassigned to a Select Stan/Eval crew and on March 15, 1972 joined the initial Bullet Shot rotation of B-52G's to Guam. In this process, then Captain Fritz completed 185 B-52 missions including a mission over Hanoi on the infamous third night and again over Haiphong on the 26th. He was awarded the Air Medal with 2 silver and 3 bronze oak leaf clusters, the Distinguished Flying Cross and the Meritorious Service Medal among others.

Then Captain Fritz was reassigned to Headquarters Strategic Air Command in April '73 in the Department of Aeronautical Requirements with responsibility for avionics development primarily for the FB-111A and IR systems.

Major Fritz was reassigned to HQ, AFSC in 1978 and almost immediately to HQ USAF in the Directorate of Research, Development and Production with emphasis as Program Element Monitor for development of Strategic Electronic Warfare Systems and for the EF-111A development and initial production. Lt. Col. Fritz was also designated as the USAF Delegate to NATO for Electronic Warfare and a member of the Air Force Board, recommending overall USAF budget allocations. Lt. Col. Fritz was active in the development of the first version of the Towed Decoy.

Lt. Col. Fritz retired in August 1983 and joined Tracor Aerospace as a Program Manger. Shortly after that he was recruited by E-Systems where he became Director of International Business Development. Mr. Fritz remained with E-Systems at it's facility in Greenville, TX, through it's purchase by Raytheon and eventually by L-3 Communications and was instrumental in the successful pursuit of several key international programs verses Boeing, Lockheed and Marshals of Cambridge including the RAAF P-3 avionics upgrade program, RNZAF C-130 self protection and later structural and avionics upgrade programs.

Bill Fritz has a Bachelor of Science from Rutgers University and a Master of Arts in Business Management from the University of Nebraska, and has completed all USAF degree equivalent schools through the Industrial College of the Armed Forces. Bill continues to advance his education and consult on a limited bases. He and wife, Cheryl, live in Dallas, TX. Both daughters graduated from SMU, were employed by the US Congress and are now in responsible positions in the FDA and RLM Marketing.

Bill is active in support of improved recognition of the Strategic Air Command and has available a SAC auto magnet and a B-52 mission tape with the accompanying music from *"The Right Stuff"* and his mission tape from the Hanoi mission on 20 Dec. 72.

Richard L. Gaines

Colonel, USAF (Ret.) Col. Gaines retired in 1983 after 27 years of service, all but 3 of which were in SAC. He piloted B-52D, B-52E, B-52F, B-52G and B-52H aircraft for over 6000 hours and the F-86 and B-47 for another thousand. He, his wife Bunny and children Kristi and Lee were stationed at McCoy AFB, Fl, Walker AFB, NM, Grand Forks AFB, ND, Castle AFB, CA, Seymour-Johnson AFB, NC, Loring AFB, ME and Dyess AFB, TX. Over those years, Col. Gaines held most Squadron and Wing positions including two years as OMS Commander at Seymour-Johnson AFB . He flew more than fifty combat missions as pilot in the B-52D during the Viet Nam Conflict. His alert pulling crew duty extended from 1959 to 1976. Among other honors, his crews won the Eisenhart Trophy for the best B-52 crew in 15th Air force and another was selected to represent Grand Forks in the 1970 Bombing and Navigation Competition at McCoy AFB, Fl. From 1980 to 1983 he was assigned as Director of Operations for the 68Th Bomb Wing at Seymour-Johnson, Director of Operations at the 42nd Bomb Wing at Loring and held the same position at the 12th Air Division at Dyess.

Charles Haigh

Major, USAF (Ret.) Born and raised in Florence, SC. Graduated College of Charleston in 1965 with a BS in biology. Worked for Celanese Fibers Co. in Charlotte, NC then joined the Air Force in September 1966 to become a pilot and kill commie bastards in Southeast Asia.

After OTS, Lt. Haigh reported to Moody AFB, Georgia, for UPT training. He graduated in February 1968 and reported to England AFB, Louisiana, for AC-47 'Spooky' gunship pilot training. He then attended Jungle Survival School at Clark AB in the Philippines, before reporting to the 3rd ACS at Bien Hoa AB, RSVN, in July 1968, for a one-year tour.

During his tour at Bien Hoa, Lt. Haigh flew approximately 225 night, low-level, close-air combat support missions, first as a copilot, then as an AC. He fired over two million rounds of 7.62 supporting friendly forces on the ground, was shot at hundreds of times without being hit, and was awarded the DFC, multiple Air Medals, and the Vietnamese Cross of Gallantry w/ Palm (unit citation).

After completing his tour, Lt. Haigh was assigned to the 5th Bomb Wing at Minot, ND, to fly the B-52H. He completed a copilot Arc Light tour out of U Tapao, then returned to Minot and checked out as an AC. Capt. Haigh and his crew checked out in the B-52D at Castle AFB, then reported to Andersen AFB, Guam, for his second Arc Light tour.

After a number of Arc Light missions, Capt. Haigh and his crew flew four missions over North Vietnam during Linebacker II. Capt. Haigh flew one of these missions with an engine shut down due to a firelight shortly after takeoff. Forced to fly lower than the rest of the cell, Capt. Haigh's plane lost mutual ECM protection and, thus, came under nearly continuous uplinks and SAM attacks during this mission. Capt. Haigh managed to avoid all the SAMs, enabling him and his crew to drop their full bomb load on their assigned target. For this mission, CINCSAC General John 'J.C' Meyer pinned the Silver Star on Capt. Haigh. The rest of his crew were awarded the DFC.

Maj. Haigh soon found that all his hero medals and a dollar would buy him a cup of coffee. A victim of the post-Vietnam RIF, he served a short stint in enlisted status at Los Angeles AFB, California, then retired honorably as a Major in October 1986. 'It all counts for 20,' he reminded himself.

After retiring from the Air Force, Charles worked as an actor (his first role on 'Little House on the Prairie') and writer (briefly writing for 'Divorce Court'), then taught high school English and ESL at Hawthorne High School in Hawthorne, California for 12 years. He retired from teaching in 2004 and became a double-dipper retiree.

Charles is a member of S.A.G. and has been acting in TV, film, and on the stage and writing novels, short stories, and his memoirs in retirement. His short list of acting credits may be viewed at www.imdb.com. He has published a short story, 'First Solo,' and co-wrote an original screenplay, 'In a Moment of Passion,' a feature film that was shot in Poland. He also worked as an actor in the film. Charles has recently completed a book, "King of the Bees," which will soon be available for purchase at www.trafford.com.

Charles lives now in Torrance, California, with his lovely Chinese wife, Francesca. He plays golf, rides his Harley Super Glide, takes power walks by the beach, travels extensively, and continues his writing and acting career. He has three adorable granddaughters and two more grandchildren on the way. Life is good!

Gary Henley

Colonel, USAF (Ret.) was commissioned in 1973 through the Texas A&M University ROTC program. His operational flight experience includes the B-52D, RC-135S (Cobra Ball), and RC-135X (Cobra Eye) aircraft. He served as a flight instructor/evaluator, crew commander, flight commander, operations officer, and squadron commander. He has an extensive background in electronic and information warfare (EW & IW) in the areas of training, systems engineering, and defensive systems flight testing. He served in various staff positions at Wing, Center, MAJCOM, and Agency levels in his career. Unique duties included Assistant Deputy Chief of the Central Security Service; Chief, M04 (National Security Agency); and Chairman of the National Emitter Intelligence Subcommittee (under the National Signals Intelligence Committee).

He earned technical specialty badges in both navigation and intelligence career fields, finishing his AF career as the vice Wing Commander of the 67th Information Operations Wing at Lackland AFB, TX, where he retired in 2003 after 30 years of service in the USAF.

Currently, he is the Director for the San Antonio office, Information Science & Engineering Center, of Syracuse Research Corporation and Senior Technical Consultant to the Technical SIGINT Airborne Program Office at the National Security Agency.

His awards include the Defense Superior Service Medal, Legion of Merit Medal, Meritorious Service Medal with four oak leaf clusters, Air Medal with one oak leaf cluster, Air Force Commendation Medal, Air Force Achievement Medal and Combat Readiness Medal with one oak leaf cluster.

Marvin W. Howell

Colonel, USAF (Ret.) Marvin W. Howell retired in September 1990 as the Director of Intelligence Systems, Deputy Chief of Staff Intelligence, Headquarters Strategic Air Command, Offutt Air Force Base, Nebraska.

Colonel Howell was born 1 January 1938 in East Alton, Illinois. The Colonel graduated from Southern Illinois University with a bachelor's degree and as a distinguished graduate of Air Force Reserve Officers Training Corps (ROTC) in 1961. He subsequently earned his master's degree from Troy State University in 1973 and a follow-on Master's Degree from UNO after retirement in 1990. Colonel Howell is a graduate of Squadron Officer's School, Air Command and Staff College, Industrial College of the Armed Forces, and the Air War College.

Colonel Howell was commissioned a second lieutenant through ROTC on 17 June 1961 and entered the Air Force at Craig AFB, Alabama the same month for pilot training. He was subsequently assigned to James Connelly AFB, Texas in 1962 where he earned navigator's wings. He was assigned to Mather AFB, California in 1963 for Electronic Warfare Officer training. His first operational assignment was to Glasgow AFB, Montana in 1964 where he accumulated over 2500 hours in the B-52D, including two Arc Light tours. He was on the first crew to land at Kadena in response to the Pueblo Crisis. The Colonel flew over 100 Arclight missions and was on the wave lead crew for over 50 missions. His crew was selected for the Top- 3 award by Third Air Division and was decorated by then General Nuygen Cao Ky during the Guam Summit of Presidents Johnson and Thieu.

Remaining in B-52's, Colonel Howell was transferred to Dyess AFB, Texas when Glasgow closed. The Dyess tour was interrupted by Squadron Officer's School and a subsequent assignment to EC-47's at Phu Cat AB, Republic of Vietnam in 1969-70. Colonel Howell flew 118 combat missions in the EC-47.

Colonel Howell returned to the CONUS in 1970, assigned to the Squadron Officer's School faculty. He was subsequently selected to attend Air Command and Staff College (ACSC) in 1974 where he finished in the upper seven percent. Following ACSC, Col Howell was assigned to the FB-111, 1970-74, at Pease AFB, New Hampshire on the wing staff.

Colonel Howell entered the Intelligence career field while assigned to Intelligence Center Pacific, 1976-79. He was subsequently assigned to the Joint Strategic Target Planning Staff (JSTPS) at Offutt AFB, Nebraska where he served as Chief of the Data Section in the National Strategic Target List (NSTL) Directorate (1979- 80) and Chief, Combat Targeting Team (1980-81).

Promoted to Colonel in 1981, he was assigned as Commander of the Strategic Target Intelligence Center (STIC). After a year in the STIC, Colonel Howell was assigned as the Assistant for Air Force Programs, General Defense Intelligence Program (GDIP). He served in that duty for 10 months before being called back to duty with the JSTPS in June 1983 as Chief of the Weapons Allocation Division in the NSTL Directorate. Colonel Howell was selected to be the Director of Targets, DCS/ Intelligence in March 1984 where he served until October 1987 when he was handpicked to form and direct a new Directorate of Intelligence Systems.

Colonel Howell's awards and decorations include Defense Superior Service Medal, Distinguished Flying Cross, Defense Meritorious Service Medal with One Oak Leaf Cluster, Air Medal with 7 Oak Leaf Clusters, Air Force Commendation Medal with 2 Oak Leaf

Clusters, Republic of Vietnam Air Medal, and Republic of Vietnam Air Gallantry Medal.

Col Howell is married to the former Anita M. Anglin of Roxana, Illinois. They have three children, Bryan and twins Damon and Matthew. He was in private practice as a Mental Health Counselor for five years and is now fully retired.

John W. (Bill) Jackson

Colonel, USAF, (Ret.) He enlisted in the aviation cadet program in August 1942. He was in the class of 43-J and was awarded pilot wings and a commission as a 2nd Lt. US Army Air Corps, November 1943. At that time as an army pilot he was told he could fly any type of plane that was available! Sure. Upon completion of B-17 phase training as a copilot, he was assigned to the 95th Bomb Group, (first B-17s over Berlin), 8th Air Force. He completed 35 missions July 1944. He participated in the air assault on D-day and the first 8th Air Force shuttle mission to the Soviet Union. He was awarded the DFC, Air Medal with 4 clusters, and the ETO medal with four battle stars.

After VJ day, he was transferred to MacDill Field, Tampa, Florida. His last assigned duty at MacDill AFB was that of a flight supply officer. His account included the base cannon. He was released from active duty January 1947 with the rank of captain. He was active in the Air Force Reserve and completed four summer active duty tours and was recalled to active duty May 1951. He was then assigned to SAC, 43rd Bomb Wing, Davis Monthan AFB, Tucson AZ. He served as a B-50 copilot for two years and then upgraded to aircraft commander July 1953.

When the B-50's were sent to the adjoining bone yard, he was transferred to the other squadron on base in the 303rd Bomb Wing, just getting combat ready in their new B-47's. He completed B-47 transitions school at Pine Castle, Florida, in March 1954. His crew was selected for upgrading to the B-52 program in October 1957. Upon completion of CCTS at Castle AFB, CA, he was transferred to the 325th Bomb Sqdn, 92nd Bomb Wing, Fairchild AFB. WA. He participated in the seven-month airborne alert test, flying 22 missions.

In January 1961, he was promoted to major, and promptly was transferred to the 326th Bomb Squadron. When SAC renamed its units for World War II units the Glasgow unit became the 91st Bomb Wing, 322nd Bomb Squadron.

He upgraded to IP in 1963, served in the standardization section and was one of two aircraft commanders sent to Guam for indoctrination in Arc Light procedures prior to the wing's scheduled six-month tour beginning in September 1966. He was promoted to Lt Col in the spring of 1966. He flew several Arc Light missions before becoming chief of the crew scheduling section.

Upon return to Glasgow, he was assigned as chief of the programs and scheduling section. He was transferred to Beale AFB, CA, in February 1968, and after attending B-52G difference upgrading at Castle AFB, he was assigned to the scheduling section while still being a member of the bomb squadron.

He was promoted to Colonel in the reserve forces in the spring of 1969. Choosing to remain on active duty as a Lt Col he transferred to the B-52 operation in Thailand in November of 1969. However, the President decreed that all reserve officers with 20 or more years of service should be retired in six months. He retired 31 March 1970 having logged more than 7,100 hours of flying time with 4,100 in the B-52B, D and G models.

He completed work towards an MS degree in Gerontology after he retired, receive his BA degree in Social Welfare and Corrections, and was over half way to an MA degree from Chico State when he decided to call it quits. He currently resides in Yuba City, CA.

Lothar "Nick" Maier

Major, USAF (Ret.) was born in BUFFalo, NY. He entered pilot training with Aviation Cadet Class 55-M, January 1954, and was commissioned at Williams AFB, Arizona, April 1955. Immediately after graduation, he was one of the first Second Lieutenants to enter SAC's Pilot AOB (Aircraft-Observer-Bombardier) course at James Connally AFB, Texas, and received a Navigator rating. Assigned to B-47s at Smoky Hill AFB, Kansas, where in 1956 his crew was the first from the 40th Bomb Wing to be assigned to B-52 upgrade training at Castle AFB, and subsequently remained there in the 93rd Bomb Wing training cadre.

Nick was a B-52 aircraft commander for twenty years, flying the B through G model aircraft. He received a SAC Crew of the Month Award for an aircraft save in 1967. Served one B-52 Arc Light tour in 1969 with 70 combat missions, and was 8th AF Senior Controller at Andersen AFB, Guam, during Linebacker II in 1972. Retiring as a Major in 1977, he worked 16 years in Travel Industry Management. He is married to Mary Beth, and their son Robert has an Instrument Rated Commercial Pilot's license.

Walt Marzec

Lt Colonel, USAF (Ret.) Graduated from the University of Connecticut and the ROTC program in 1970, with a BA in "Do you want fries with that?" English. Went to Nav & EW schools and ended up at Fairchild AFB, WA. After upgrading in the B-52G at Ellsworth AFB, went to Castle AFB for "D" model cross training (cross dressing) and then overseas to fly 42 combat missions out of Guam and Thailand. Spent only two years in BUFFs (1972-1974). In 1974 got a Palace Cobra assignment to F-4s. Did five years between Homestead AFB, FL; Clark AB, PI; and RAF Bentwaters, UK. Got another assignment to F-111s and did four years at Mountain Home, ID and RAF Upper Heyford, UK. Left flying and went to HQ TAC, NSA, AFEWC, and finally an OSD project. Presently, running the Fryolator at McDonald's...no, only kidding! Actually, now working for Syracuse Research Corp., San Antonio, as a Program Manager.

Robert Gary Miller

Former Captain, USAF, (Honorably Discharged) Gary was born into a military family in Seattle Washington in 1944, the son of a naval officer. He grew up in the Navy, moving with his family from station to station across the United States and Europe. Upon graduation from New Mexico State University, and being commissioned a Second Lieutenant in the U. S. Air Force, he attended Navigation Training and then Bombardier training at Mather AFB in California. Following flight training, he was assigned to the 668th Bombardment Squadron, Strategic Air Command (SAC) where he flew B-52G Stratofortress aircraft, aiding in SAC's global "Cold War" mission.

To provide global responsiveness with conventional weapons in Southeast Asia, he flew "Arc Light" missions out of U-Tapao, Royal Air Force Base Thailand in 1970-71. After returning to civilian life, he obtained two Masters Degrees – one in Educational Management and the other in Fine Arts - from New Mexico State University.

In 1976 Mr. Miller was engaged by the CEO of Atlantic Richfield Oil Company (ARCO), to serve as Director for the Lincoln County Heritage Trust, a non-profit corporation in New Mexico. In 1994 Mr. Miller utilized his museum expertise and leadership skills to develop and design a new 100,000 square foot, $13,000,000 military air museum which would be dedicated to the famous 8th Air Force of WWII fame, located in Savannah, Georgia.

Mr. Miller has over 30 years experience directing various museum operations having worked for and with top military officers and Fortune 500 CEO's on a number of major projects, many of which involved creating a museum from the ground up.

At Present he is serving the U. S. Air Force as the Museum / Heritage Director of the 15th Airlift Wing at Hickam AFB, Hawaii. As Heritage Director he serves as the principal advisor to the 15th Airlift Wing commanding officer on all matters pertaining to museums, exhibits, historical aircraft, and heritage programs.

Arthur Craig Mizner

Major, USAF (Ret.) A B-52 Combat Veteran Command Instructor Pilot of Vietnam (1969 - 1973) Arc Light, Linebacker I and II missions, and a B-47/B-52 Veteran Pilot of the Cold War ground and airborne nuclear alerts. In addition, Craig flew five 24 hour airborne nuclear alert missions during the Cuban Missile Crisis, the military-political confrontation between the USA and USSR in 1962. Craig entered USAF active duty in 1954 as an Aviation Cadet (class 56Q) and was commissioned a 2nd Lieutenant and Jet Fighter Pilot on 28 June 1956 and retired 1 January 1977 with 264 TAC and SAC combat missions, 9206.7 hours in the B-52 and over 11,000 military flight hours. In addition, Craig has over 2,000 hours as a commercial pilot. Craig joined General Dynamics/Lockheed Martin Aeronautics in 1979 and was the F-16 Pilot Flight Manual Manager for the first 17 years. Since than, Craig using his computer skills works as an Advance Technology System Integrator Aerospace Staff Avionic Engineer in support of the F-16 Fighting Falcon. Craig has BS in Industrial Technology from Texas A&M – Commerce, TX December 1978 and an AA in Social Science from Tarrant Country College, TX May 1998.

.Craig was inducted in the Phi Theta Kappa International Honor Society on 20 April 1998. For Craig's involvement in the bombing of Hanoi, NVN for the return of all POWs, read "The Eleven Days of Christmas: America's Last Vietnam Battle" by Marshall L. Michel. In May 2002, Craig and his copilot Donald Allen Craig were interviewed and video taped for a TV feature on the bombing of Hanoi. The TV feature should air in the near future. See URL for details: http://www.teleproductiongroup.com/12_72-main.html Be sure to navigate to all areas. "Those Who Lived it." The first picture you will see is below. Look at the fifth interview. Craig is a contributing author to the November 2006 Smithsonian Air & Space magazine article Cuban Missile Crisis 27 days at DEFCON 2. In addition, Craig has authored many articles for We Were Crewdogs, a volume of books dedicated to The B-52 Collection as edited by Tommy Towery. In August 2007, Craig did a 3 hour video interview with the National Geographic Channel Towers Productions for a 3 hour video titled Inside the Vietnam War. The first showing was on February 18th 8:00 PM EST 2008. Craig is still working full time and now has over 29 years with Lockheed Martin Aeronautics located in Fort Worth, TX as an aerospace avionic staff engineer in support of the F-16.

Larry Moeller

Captain, AF RES Larry Moeller graduated in UPT Class 66F (Undergraduate Pilot Training) at Vance AFB. He grew up in a rural NE Iowa town of 500, is an Eagle Scout and was high school valedictorian. He learned auto mechanics at his dad's dealership and worked his way through college on that skill. He graduated as an AFROTC 2nd Lt with a B.S. in Industrial Technology from Iowa State University in 1964.

Larry served in B-52's at Amarillo & Beale before leaving active USAF in March 1970 with a hiring offer from United Airlines in hand, unaware that the economy was headed into recession and airline jobs were headed for the layoff bin. While waiting for a UAL class date, he finished a Master's in Education at the Chapman College Center at Beale, flew C-124's with the AFRES out of McClellan, and taught math & science in Roseville, CA. As airline prospects dimmed, he enrolled in a doctoral program at Iowa State in 1971 with a graduate research assistantship.

In 1973 Larry accepted a professorship in Industrial Education & Technology at Oregon State University, where he remained 5 years before joining his wife, Bunnie, in opening a real estate brokerage. Larry completed his dissertation and received the Ph.D. in Education from ISU in 1980. In 1981, divorced and remarried, Larry was recruited to a professorship at Iowa State.

In 1985 he became a community college administrator, eventually working his way back to the sierra foothills of California that he'd come to love via the USAF-induced transplant to Beale. He now lives on ridgetop acreage in the Horneblende Mountains near Georgetown, CA.

Over the years, Larry has served as a Scout leader, raised and trained 30 Arabian horses, ridden 50-and-100-mile endurance races, and enjoyed owning a Citabria taildragger. His current passion is vintage auto racing in a Porsche 914-6 and an MGB.

Since 1994, Larry has been at Gavilan College in Gilroy, CA as tenured faculty member. He's served 4 yrs as president of the faculty union. Since 1998, Larry has been the college's Grants & Development officer, earning his keep by winning over $10 million in Federal & State grants.

All-in-all, it's been an interesting ride through life. Larry plans to retire comfortably in 2009 with 32 years credit in the teacher's retirement system, and he is no longer sorry he missed the glorious airline career and glad he isn't holding an empty bag like his airline retiree buddies.

Larry says that the USAF and the guys he crewed with, and those who mentored and instructed him, should be credited for imbuing this one naive young man with a philosophy that combined risk taking fighter instincts, officer-and-gentlemen honor, and comrade in arms loyalty.

John Morykwas

Former Captain, USAF, (Honorably Discharged) Active duty Navigator in the 20th BS at Carswell AFB, Texas from 1974 to 1979. Reserve Weapon System Officer (fast recon) in the 1980s. Three Honarable Discharges from the military and retired US Government.

Current occupation is Clinical Laboratory Scientist, Medical Technologist ASCP.

Karl D. "Ned" Nedela

Senior Master Sergeant, USAF (Ret.) was born and raised in Crete, Nebraska. After high school, attended Doane College in Crete and played 2 years of football. In 1951, I knew I was about to be drafted (not by the NFL) so I enlisted in the Air Force.

After a year in Electronics Courses at Lowry AFB, I attended B-36 Gunnery School in Denver. I went through B-36 Transition at Carswell AFB. Our group was sent to Loring AFB and I served as an instructor gunner for 3 years. In 1956, I was sent to B-52 Gunnery School at Lowry AFB and then to Castle AFB. Returned to Loring and our Squadron was sent to Biggs AFB, El Paso, in the Summer of 1959.

Reassigned to Dyess AFB, Abilene, Texas in 1966. In June of 1970, I went for seven months to U-TAPAO as a TDY Wing Gunner.

I graduated from McMurry College in 1972, thanks to the Bootstrap Program. Graduated in August and in September went as a crew gunner to U-TAPAO. I returned in April of 1973 and retired in July of 1973. We moved to Killeen, Texas where I taught and coached for 23 years.

I have been married to my wonderful wife, Elizabeth, for 53 years and have 2 children and 3 grand children. I have fond memories of my Air Force career.

Gery Putnam

Lt Colonel, USAF (Ret.) Commissioned USAF from Naval Academy in 1957. Following pilot training and B-47 crew training assigned to 337th BS, 96 BW, Dyess AFB, TX. 1962-1964 Minuteman launch crew member. Back to Dyess and the 337th in B-52E. Arc Light Jan - Jun 1969. AFSC completed Jan 1970 followed by SEA tour flying C-7A. 1971 - 1974 DOTTA at 15th AF HQ. Naval War College completed in July 1975 followed by three year tour as staff at AFSC and retirement as Lt Col in Aug 1978.

Goofed off after that mainly teaching recreational scuba diving. Since 1997 aquatic instructor and lifeguard at rec center in Virginia Beach, VA. Separated, one son, one daughter, both grown.

Wade Robert

Lt Colonel, USAF (Ret.) earned a B.S Degree in Business Administration in 1967 and attended Air Force Officer Training School at Lackland AFB, TX and was commissioned a 2nd Lieutenant in February 1967. He then was assigned to Lackland Air Force Base, TX as a personnel officer. Wade entered pilot training at Randolph Air Force Base, TX winning his wings in August 1969 and then attended B-52 Combat Crew Training at Castle AFB, CA. Upon completion of the training he was assigned to the 486th Bomb Squadron and later the 2nd Bomb Squadron, 22nd Bomb Wing at March AFB, CA flying B-52C, B-52D and B-52F. He completed one Arclight deployment as a copilot, upgrading to aircraft commander shortly after his return. He and is crew were on constant deployment with only 48 day leaves until the summer of 1972, having participated in Operation Bullet Shot and Linebacker II and flying over 200 Arclight missions. He returned to March AFB as an instructor pilot and bomber scheduler until his assignment to the Concepts Directorate at Strategic Air Command (SAC) Headquarters where he helped develop new nuclear mission concepts. From SAC he was assigned to the Pentagon serving in multiple roles including the B-1 Program Element Monitor in Research and Development and in the Programs Division of Air Force Legislative Liaison. Following his duty in Legislative Liaison he became Director of Special Projects at Air Force Systems Command working on then highly classified programs, including development of the B-2, and the F-117. His final assignment was as Special Assistant to the Assistant Secretary of the Air Force for Research and Acquisition. Wade retired in February 1987 after a 20 year career.

His decorations include the Legion of Merit, Distinguished Flying Cross with oak leaf cluster, Meritorious Service Medal with oak leaf cluster, Air Medal with two silver and one bronze leaf cluster, Air Force Commendation Medal with one oak leaf cluster, Air Force Outstanding Unit Award with Valor and silver and bronze oak leaf clusters, Air Force Organizational Excellence Award with oak leaf cluster, Combat Readiness Medal, National Defense Service Medal, Armed Forces Expeditionary Medal, Vietnam Service Medal with two bronze stars, Republic of Vietnam Campaign Medal and the South Vietnam Cross of Gallantry with Palm.

After retirement from the Air Force he joined United Technologies Corporation (UTC) where he served for 15 years working with the Congress, The White House and various departments and agencies as Director and then Vice President of Government Affairs in Washington, D.C. retired from UTC in 2002 and lives with his wife in Alexandria, Virginia and Sanibel, Florida.

Glenn Russell

Former Captain, USAF (Honorably Discharged) Glenn was raised in the small town of Pleasant Hill, Missouri, and went to college at Central Missouri State College which is now known as University of Central Missouri. After graduating in 1965 it was find a job that would give me a deferment, get drafted, or join up. I talked to the Air Force Recruiter and was accepted into pilot training. That sounded like a better option than being drafted. After Pilot training I was assigned to the F-4 as a Pilot Systems Operator. After a year and a half in the F-4 I was assigned to SEA as a Forward Air controller flying the O-1 Bird Dog. I spent a year in Vietnam supporting the 173rd Airborne Brigade.

I was assigned next to the B-52 G at Fairchild AFB. In early 1972 one of the crews was preparing to go Arc Light and the Crew commander was injured and unable to go. I was about due for an Arc Light tour so I volunteered to take the crew. As we were out processing to go to RTU some of the G-Model crews were processing out to go to Guam as the first Bullet Shot crews. So, I flew D-Models out of Guam and U-Tapao until the end of 1973. My crew flew six missions during Linebacker II. I left the Air force in 1973 and spent the next 25 years flying with the airlines.

Kenneth R. "Ken" Schmidt

Major, USAF (Ret.) was born in Shattuck, Oklahoma, grew up in Woodward, Oklahoma, and earned his B.S. degree in Business Administration from Southwestern (Oklahoma) State College in 1970. He entered the USAF through Air Force Officers Training School in April, 1974. He was then assigned to Mather AFB, CA, for Undergraduate Navigator Training (UNT) where he received training in the T-29 and T-43 aircraft. Upon graduation from UNT and earning his navigator wings in December, 1974, Ken continued his training at Mather completing the ASQ-48 Navigator/Bombardier Training (NBT) course in April, 1975, and was assigned to the 9th Bomb Squadron, 7th Bomb Wing, at Carswell AFB, TX. He completed Combat Crew Training School (CCTS) and was assigned to a combat crew. At Carswell, he served as navigator, instructor navigator, and radar navigator in the B-52D and left Carswell in 1980. His next assignment was to the 23rd Bomb Squadron, 5th Bomb Wing at Minot AFB, ND, flying in the B-52H. While at Minot, he served on a combat crew as a radar navigator and instructor. He was later assigned to the 5th BW as a Bomber Scheduler. In 1983, Ken was transferred to Castle AFB, CA, as an Academic Instructor at the 4018th CCTS and served in that position for two years. His flying at Castle was in the B-52G. He was then assigned as the Chief, Air Weapons Branch, and remained in that position until being transferred in 1987. His next assignment was to Offutt AFB, NE, and he was assigned to the 2nd Airborne Command and Control Squadron (2nd ACCS) as an Operations Plans Officer flying onboard the EC-135C "Looking Glass". In 1989, Ken was assigned to HQ/SAC DOO and held various staff jobs in the DO community until SAC closed up shop in 1992. He was then reassigned to the 55th Strategic Recon Wing and worked in the Wing Inspector's Office until his retirement in 1993. He accumulated 3651 flying hours: 155 hours in T-29/T-43 trainers; 2681 in the B-52D, G, and H; and 811 hours (100 operational missions) in the EC-135C.

While in the Air Force, Ken completed his Master's Degree from Texas Christian University. He also completed Squadron Officer's School (by correspondence and in residence), Air Command and Staff College, and Air War College.

Ken remained in Papillion, Nebraska, after his retirement and is currently an assistant director in the Financial Aid Office at the University of Nebraska at Omaha.

Ken was married to Susan (Datin) for 34 years and has two daughters, Stephanie and Staci. Susan passed away on 1 July, 2004, after a 5-year battle with breast cancer. On December 16, 2006, Ken married Nancy Schuler. Formerly from Sioux Falls, South Dakota, Nancy moved to Omaha in 2005. Nancy has two sons, Ryan, 23 and Chris, 26.

Pete Seberger

Major, USAF (Ret.) Pete Seberger graduated from the University of Nebraska in June 1962 with a BSME and a commission as a 2nd Lt. from AFROTC. He entered pilot training at Vance AFB that November and graduated in December of 1963 with class 64-D. He was assigned to SAC at Barksdale AFB to fly B-52Fs and after CCTS, survival, and nuclear weapons schools he began crew life in June of 1964. His unit was sent TDY to Guam in early 1965 and was transferred to Carswell AFB in May-June of that year. The combined wing then flew most of the early Arc Light missions until December of 1965 and in mid 1966 the D model fleet took over those duties. In April of 1968 the Carswell unit was reduced to one bomb squadron and about half the personnel were dispersed to other units in SAC. Pete went to Grand Forks AFB, upgraded to AC and flew two RTU tours before upgrading to IP. In 1973 he was assigned to Castle AFB as a flight line instructor for most of that tour. In 1977 he was assigned to 13th. AF at Clark AB where he filled the remote tour square in several different jobs. He was reassigned to Ellsworth AFB (SAC never lets go) as a crew dog, requalified as an AC and IP simultaneously, and in June of 1979 was selected to command the Physiological Training Unit at Ellsworth, while remaining as an attached instructor to the two squadron bomb wing. (Yeah! No alert but lots of flying!) In early1983 the unit was reduced to one bomb squadron so he went off flying status. He retired in December, having logged 6600 hours flight time in the B-52, some of it every year between 1964 and 1983.

After his military service he worked for a small flight operation in Rapid City and earned his ATP and civilian instructor ratings, flew a year or so for a small airline, and finished his flying career as a Flight Safety instructor in the Beechcraft King Air series at the factory school in Wichita. His logbook is just shy of 9000 total flight hours, over 4000 as instructor.

Kenneth B. Sampson

Captain, USAF (Ret.) I studied Aircraft and Powerplant mechanics in high school and became a licensed A&P mechanic in 1957.

I was an enlisted jet engine mechanic in USAF for two years and an enlisted USAF Cadet Candidate at the West Point Preparatory

School for one year. I was a dual status A/2C and Air Force Cadet at the USAF Academy for four years graduating in 1964.

I tried pilot training but was not physically coordinated enough to be a pilot, so I became a B-52D Navigator and Radar Navigator. I flew a total of 363 B-52D combat missions. I flew 343 B-52D combat missions bombing South Vietnam, Laos, and Cambodia. I flew 20 combat missions over North Vietnam. I flew 312 combat missions as a navigator and 51 combat missions as a radar navigator. I flew zero missions in Linebacker II, because I was at Carswell AFB for upgrade training when Linebacker II hit.

As a USAF B-52 navigator / bombardier, I exercised my responsibility to make an effort to win the war and free the prisoners (of which 10 were my USAFA classmates) by aiming and dropping over 35,000 bombs, with the intention of killing as many of the enemy as I could with each bomb.

I was awarded one DFC, 18 Air Medals and a 300-mission patch. I flew 2205.7 combat flying hours and 3945.3 total flying hours in the B-52D in seven years. I was stationed at Amarillo, Homestead, March, and Dyess. I had eight Arc Light tours, one year on Guam, and two years at U-Tapao and Okinawa. I left U-Tapao in November 1973 for five months at Sheppard AFB hospital and was medically retired as a Captain for a nervous breakdown.

I have been married to my Thai wife for 33 years. We have between us five children, six grandchildren, and one great grand child.

E.G. "Buck" Shuler Jr

Lt General, USAF (Ret.) E.G. "Buck" Shuler Jr. commanded the Strategic Air Command's 8th Air Force, Barksdale Air Force Base, Louisiana from 26 March 1988 to 22 May 1991. The command of more than 57,000 personnel was responsible for Strategic Air Command (SAC) operations in the Eastern half of the United States, Europe and the Middle East. It comprised over half of SAC's long-range force of manned bombers, tankers and intercontinental ballistic missles. During Operation JUST CAUSE in Panama and Operation DESERT SHIELD/STORM in the Persian Gulf War, Eighth Air Force B-52G bomber, KC-135 and KC-10 tanker and TR-1 reconnaissance units made significant contributions to the successful outcome of these conflicts. During this time General Shuler actively led his units both from the headquarters and in the field flying tactical air refueling missions and two combat support air refueling missions with his troops.

General Shuler was born 6 December 1936 in Raleigh, North Carolina. He was educated in the city schools of Caracas, Venezuela and Orangeburg, South Carolina, graduating from Orangeburg High School in 1955. The general earned a bachelor of science degree in civil engineering from The Citadel, the Military College of South Carolina in 1959 and a master of science degree in management from Rensselaer Polytechnic Institute in 1967. He completed Squadron Officer School in 1964, Command and Staff Course of the Naval War College in 1972 and the National War College in 1976, all in residence.

The general, a distinguished graduate of the Reserve Officer Training Corps program, was commissioned as a second lieutenant on 6 June 1959 and entered active duty on 18 July. He completed preflight training at Lackland Air Force Base, Texas and primary flight training at Moore Air Base, Texas. Upon graduating from T-33 jet basic training at Laredo Air Force Base, Texas, he received his pilot wings on 2 September 1960. General Shuler was then assigned to the 9th Bombardment Squadron, Carswell Air Force Base, Texas as a B-52F pilot. From September 1963 until June 1966 he served with the 337th Bombardment Squadron, Dyess Air Force Base, Texas as a B-52E pilot and aircraft commander of a lead crew. During these two tours he flew 22 nuclear airborne alert missions of 24 to 25 hours duration in the B-52, six of which were flown during the 1962 Cuban Missile Crisis.

After completing his master's degree in June 1967, General Shuler was assigned to the 68th Tactical Fighter Squadron, George Air Force Base, California, as a replacement training unit fighter pilot in the F-4D Phantom II. In March 1968 he transferred to the 558th Tactical Fighter Squadron, Cam Ranh Bay Air Base, South Vietnam as an F-4C aircraft commander. He served as an assistant flight commander and flew 107 combat missions over North Vietnam, the Republic of Vietnam and Laos. He also participated in the 558th Tactical Fighter Squadron's operational deployment to South Korea during the USS Pueblo crisis, flying 15 combat support missions along the Korean demilitarized zone as well as 57 training missions in the F-4C.

Upon returning to the United States in March 1969, General Shuler was assigned to Headquarters 2nd Air Force, Barksdale Air

Force Base, Louisiana as an industrial engineer and subsequently served as Deputy Chief of the Engineering Management Division. The general completed the Base Civil Engineer Course at Wright-Patterson Air Force Base, Ohio as a distinguished graduate in November 1969. He completed the Central Flight Instructor Course at Castle Air Force Base, California in April 1970 and assumed additional duties as a T-39 instructor pilot with the 2nd Bombardment Wing standardization-evaluation board. He concluded his tour of duty at Barksdale in July 1971 as the Assistant Deputy Chief of Staff for Civil Engineering.

General Shuler graduated from the Command and Staff Course of the Naval War College, Newport, Rhode Island in June 1972 and was assigned as assistant executive officer to General David C. Jones, the Commander in Chief, U.S. Air Forces in Europe, Lindsey Air Station, Wiesbaden, West Germany. He moved with the headquarters to Ramstein Air Base, West Germany in 1973.

From July 1973 to July 1975 he served as Base Civil Engineer and Commander of the 86th Civil Engineering Squadron at Ramstein. The general completed the National War College at Fort McNair, Washington, D.C. in June 1976 and then was assigned to Offutt Air Force Base, Nebraska, where he served as Director of Operations, 3902nd Air Base Wing and as Commander of the 3902nd Operations Squadron.

In November 1976 General Shuler transferred to Headquarters Strategic Air Command at Offutt and served as Director of Programs for Engineering and Services. He was selected in July 1977 to be the Executive to General Richard H. Ellis, the SAC Commander in Chief. In April 1979 he transferred to the 19th Bombardment Wing, Robins Air Force Base, Georgia as Vice Commander and on 16 January 1980 assumed command of the wing. In July 1980 he was assigned as commander, 42nd Bombardment Wing, Loring Air Force Base, Maine. General Shuler served as commander of 4th Air Division, F.E. Warren Air Force Base, Wyoming from September 1981 until July 1984, when he then became commander of 3rd Air Division, Andersen Air Force Base, Guam. In July 1986 he returned to Headquarters Strategic Air Command as Assistant Deputy Chief of Staff for Operations and in December 1986 became Deputy Chief of Staff for Operations. General Shuler was promoted to lieutenant general 23 March 1988 with the same date of rank and assumed command of Eighth Air Force on 26 March.

The general is a command pilot with more than 7,650 flying hours, including 209 combat hours accrued during the Southeast Asia conflict. During the ten plus years as a SAC general officer, General Shuler flew 358 missions as the Airborne Emergency Action Officer on "Looking Glass", SAC's EC-135 airborne command post. General Shuler's military awards and decorations include the Distinguished Service Medal with oak leaf cluster, Legion of Merit with oak leaf cluster, Distinguished Flying Cross, Air Medal with five oak leaf clusters, Air Force Commendation Medal with oak leaf cluster, the Republic of Korea Order of National Security Merit Cheonsu Medal and sixteen other decorations and ribbons. He wears the Master Missile Badge. General Shuler is a member of Tau Beta Pi, a national engineering honor society. He received the Daughters of the American Revolution Medal of Honor in 2005 and the Distinguished Eagle Scout Award in 2006. Also in 2006, General Shuler was selected as a Distinguished Graduate of The Citadel.

General Shuler is married to Annette Fontaine Maury of Mobile, Alabama and they reside at their home in Columbia, South Carolina. The couple have three sons and eight grandchildren. General Shuler is the past Chief Executive Officer and Chairman of the Board of Trustees of The Mighty Eighth Air Force Museum located near Savannah, Georgia and has been active in a variety of community and church affairs.

Gregory C. Smith

Colonel, USAF Greg Smith is the commander of the 453rd Electronic Warfare Squadron at the Air Force Information Operations Center, Lackland AFB, TX. He is responsible all DoD Electronic Warfare Integrated Reprogramming Flagging and all Air Force "blue system" parametric data production, signatures advancement and mission planning data development as well as radar modeling and simulation for combat operations support. His squadron also supplies threat simulation to the Air Force's Distributed Mission Operations Network and Electronic Warfare Officers for deployed combat operations. Col Smith, a Henning, Minnesota native, entered the Air Force in 1987 after completing the Reserve Officer's Training Corps program at Bethel College in St. Paul, MN where he received his B.A. and North Dakota State University in Fargo, ND where he received his B.S. in Civil Engineering. He completed his M.S. in Administrative

Management from Central Michigan University, Mt Pleasant, MI in 1996. He started his operational flying career in the B-52H at K. I. Sawyer AFB, MI. In 1995, he joined the initial cadre of aircrew to fly the B-2 Stealth Bomber at Whiteman AFB, MO. His is a command pilot with over 4,000 total flying hours in the T-37, T-38, B-52G, B-52H, and B-2A. He will begin his new duties in August 2008 as the Deputy Director of Operations for the Air Force's newest major command, Air Force Cyber Command. He and his wife, Mary Jo, reside in San Antonio, TX with their children, Rachel and Nathan.

Roland E. Speckman

Colonel, USAF (Ret.) is a command pilot with 30 years of service. Col. Speckman flew 23 different aircraft in three foreign actions logging 3,300 hours of combat flying with a retirement total of 8,600 hours.

While attending the University of Wisconsin in 1941, he was accepted by the USAAC as a cadet; he graduated and was commissioned a second lieutenant assigned to the Air Transport Command. Rated a multi-engine pilot, he flew unarmed transports over "The Hump" in the China, Burma, India Theater.

Col. Speckman separated in 1945 and served as a corporate pilot for two companies in Wisconsin. He moved to Arizona in 1949 to work for Phelps Dodge Corp. in Clarkdale. Was recalled to active service in 1952 and assigned to a B-29 combat crew in the Strategic Air Command where he served as an aircraft commander for 13 years flying three different bombers, including the B-47 and B-52. Upholding SAC's motto "Peace Is Our Profession" he attended Air Staff and Command School during the Cuban crisis.

A volunteer pilot for duty in SE Asia (Vietnam), Col Speckman served two consecutive tours, the first in Recon with the Antique Airlines and then a year with the Rescue Service in helicopters. He commanded two Rescue Detachments under the Military Airlift Command. He retired from Eglin AFB in 1972.

Military awards and decorations totaled 30 to include two DFCs, 17 Air Medals and he was credited with 13 lives saved while flying with MAC's Rescue Service.

Married 64 years to Mary and has four married children, 12 grandchildren and six great-grandchildren.

Robert B. Stewart Jr.

Lt Colonel USAF (Ret.) AB and MA degrees in Education Admin & Supervision, University of Kentucky, 1946-1951. After teaching for one year, he was facing a draft call and enlisted USAF for 4 years, August 1951. After completing basic training he was assigned as a technical instructor at Lowry AFB, CO, in Officer Personnel and Career Guidance School. Discharged August 1952 to accept reserve commission with direct appointment and no break in service. Completed navigation flight training at Ellington AFB, TX., 1953. Completed ECM flight training at Keesler AFB, MS, 1953. Assigned to 67 Tactical Recon Wing, K-14, Kimpo AB, Korea, January 1954 and 1st Marine Air Wing. Assigned to 548th Recon Tech Sq, Tachikawa, Japan, July 1954. Assigned to Hq FEAF, Tokyo, Japan, July 1954 with RC-135 mission planning and ECM intelligence collection duties. Assigned to Loring AFB, ME, as Wing OJT Coordinator., 1956. He became a victim of rheumatoid arthritis and was hospitalized at Chelsea Navy Hospital and Walter Reed Hospital for six months. At this time he elected to remain on active duty. He volunteered for assignment to Keesler Tech Training Center, MS, 1957 as ECM classroom and flight instructor, intelligence librarian and as a briefer on a traveling USAF ECM familiarization team. One of two officers selected from 15,000 assigned to ATC to receive a regular commisson . Assigned Glasgow AFB, MT, with B-52 crew duties as IEW and standardization evaluator from 1957-1962. This included a six-month TDY Arc Light assignment to Guam. Then he was surprised with an assignment to Hq 15AF, March AFB, CA, as staff officer for EW training and operations, 1967-1971. Assigned to Dyess AFB, TX, as 96 Bomb Wg, Chief of Penetration Aids, 1971-1975. During this time he was TDY to U-Tapao AB, Thailand, as B-52 mission briefer. Upon returning he augmented the crew training staff at Carswell AFB. Retired September 1975 with 24 years of continuous service. Logged 5,654 flying hours, mostly in B-52 aircraft, with 630 hours of combat time. Logged ECM time in B-26, C-45, C-47, AD-4N, KC-97, C-54D, T-39 and B-52D. Citations include The Meritorious Service Medal, Commendation Ribbons, Air Medal with one Oak Leaf Custer, National Defense Service Medal, Korean Service Medal, United Nations Service Medal, Air Force Outstanding Unit Citation Award with one OLC, Air Force Longevity Service Award with 4 OLC, Good

Conduct Medal, and Vietnam Service Medal. Currently resides with his wife, Jean, in Versailles, KY, near Lexington.

George R. Thatcher, Ed.D

Major, USAF (Ret.) George Thatcher joined the Air Force in 1951, soon after graduating from his hometown high school in West Orange, New Jersey. He spent five years as an administrative specialist and training NCO, then attended and graduated from Air Force OCS Class 1957-B. After two years as an Air Defense Weapons Director, he was accepted into pilot training and graduated in Class 1960-H.

George chose to fly the "heavies" and was assigned as a B-52 copilot at Turner AFB, Georgia, where he spent over six years, upgrading to Aircraft Commander along the way. He flew many airborne alert missions, especially during the Cuban missile crisis, and had eight months of combat flying out of Anderson AFB, Guam. When Turner AFB closed in 1967 he was assigned as Aircraft Commander and Instructor Pilot to Plattsburgh AFB, New York.

In 1969 he volunteered for another tour of duty in Southeast Asia, this time flying the EB-66 "Destroyer" at Takhli RTAFB, Thailand. Following that experience, he spent his final two years of active duty as a Command and Control Officer at Blytheville AFB, Arkansas.

Since retiring from the Air Force in 1972, George has been a real estate broker, mortgage loan officer, middle/high school Spanish teacher, prison educator, and university professor of Education. He is retired from Texas Tech University and presently serves as Adjunct Professor of Workforce Education for Southern Illinois University's off-campus degree program. His major passion is writing/collecting aviation poetry and short stories, and he engages in a variety of fitness activities. He gives up the game of golf frequently.

George is married to Jean Schwisow-Thatcher, his Beauteous Wife and Princess for Life, whom he credits with teaching him everything he knows about the Education profession.

Art Thompson

Lt Colonel, USAF (Ret.), was born and raised in Gary, Indiana. After earning his BS Degree from Purdue University, West Lafayette,

Indiana, in 1963, he entered the U.S. Air Force Officer Training School at Lackland AFB, San Antonio, Texas, in February, 1964. Upon receiving his Commission in May, 1964, he went on to Undergraduate Navigator Training at James Connally AFB, Waco, Texas, where he received his Navigator Wings in May, 1965. He then entered Navigator/Bombardier Training at Mather AFB, Sacramento, California, then on to B-52 Combat Crew Training at Castle AFB, Merced, California.

Lt/Col Thompson was then assigned to the 379th Bomb Wing (B-52H), Wurtsmith AFB, Oscoda, Michigan, where he reported in April, 1966. He was EWO Certified as a B-52H Navigator on June 1, 1966, at which time was immediately assigned to Crew E-38 and placed on nuclear ground alert. During the next few years he pulled his share of training sorties, nuclear ground alert, and nuclear air-born alert "Hard Head" missions over Thule AFB, Greenland.

In September, 1968, Lt/Col Thompson was deployed on a six-month Arc-Light tour as the Navigator on the third Wurtsmith crew to be deployed. Following a two-week "D Difference" School at Castle AFB for H-Model crews, he flew a total of 50 combat missions over Vietnam from Andersen AFB, Guam, Kadena AB, Okinawa, and U-Tapao AB, Thailand.

Upon return from Southeast Asia, Lt/Col Thompson voluntarily separated from the active duty Air Force with an Honorable Discharge, and was subsequently accepted into the Wisconsin Air National Guard, Milwaukee, Wisconsin, in May, 1970, as a KC-97 Navigator (Traditional Guardsman). At about the same time, he also joined the Oscar Mayer Foods Corporation in Madison, Wisconsin, as a "full time" Manufacturing Project Engineer.

During the next 20 "some-odd" years, he juggled "masquerading" as a full-time Project Engineer and a "part-time" Guardsman. Many times the "part-time" Guardsman status turned into a "full-time" status, including many two-week TDY's to far-off places. He served in the Guard as a KC-97 Navigator for seven years, and KC-135 Navigator for 15 years, retiring in 1992 with 28 years total military service and 5500 hours military flying time in B-52's, KC-97's, and KC-135's. The last five years in the Guard were served as Squadron Chief Navigator, including a four-month TDY in Desert Storm, flying 31 Combat Support Missions out of Jeddah, Saudi Arabia, and Cairo West, Egypt.

One side note: While in Cairo West during Desert Storm, he flew a KC-135 Air Refueling Mission with a Wurtsmith BUF that was returning from a strike mission over Baghdad. (In this war he was at the other end of the boom). As it turned out, a few hours later he met that same BUF crew in "Tent City" at Cairo West, as the BUF was not able to take on the required offload to enable it to RTB. So the BUF crew had to make an emergency landing at Cairo West. Comment: "WOW....That Wurtsmith BUF crew looked a lot younger than when I was there 20 years ago."

In 1996, Lt/Col Thompson retired from the "Hot Dog" Factory with 26 years service, after which he started his own Engineering Consulting Business. He "retired" from the consulting business in June, 2007 (third and final retirement). He and his wife, Pat, a retired RN (Registered Nurse, not Radar Navigator), are both retired and reside in Sun Prairie, Wisconsin. Life is Good!.........There IS life after work!

Tommy Towery

Major, USAF (Ret.) earned a B.S. degree in Journalism in 1968 from Memphis State University where he also earned a commission as a 2nd Lt. through the AFROTC program. Attended Navigator Training and Electronic Warfare Officer Training at Mather AFB, CA. Following B-52 CCTS training in B-52F models at Castle AFB, CA he was assigned to the 20th Bomb Squadron at Carswell AFB, TX flying B-52C and B-52D models. He served as a 7th Bomb Wing Combat Intelligence briefer and a Penetration Aids staff officer while grounded from flying status for kidney stones. He was deployed for six months to Guam as part of "Operation Bullet Shot" and was assigned to 8th Air Force Bomber Operations as an Arc Light mission planner. Three month after returning from his first deployment he was sent on a second six-month TDY to Guam where he worked as an Arc Light planner in the 43rd Bomb Wing Bomber Operations. It was during this assignment that he flew on B-52 combat missions as a staff officer and helped plan Linebacker II missions. Upon return to full flight duty status he was assigned to a B-52D crew at Carswell AFB, TX. Shortly thereafter he deployed to Guam and Thailand with his first crew and later progressed from crew duty to instructor to standboard duty on Crew S-1. In 1976 he and his crew received the Mathis Trophy, awarded to the top bomber unit based on combined results in bombing and navigation in the SAC Bomb-Nav Competition.

Following his B-52 assignment he was transferred to the RC-135 program at Offutt AFB, NE and flew operation reconnaissance missions from forward operating bases. In 1983 he was assigned to a four-year tour as an Electronic Warfare Intelligence officer and briefer at RAF Mildenhall, UK. Upon his return to the states he spent his last year of active duty as an Electronic Warfare Intelligence officer with the 55th Strategic Reconnaissance Wing at Offutt AFB, NE. During his career he logged over 1,600 hours in B-52s and over 5,000 hours total flight time. His decorations include the Meritorious Service Medal with oak leaf cluster, the Air Medal with eight oak leaf clusters, the Armed Forces Expeditionary Service Medal, the Republic of Vietnam Campaign Medal and the South Vietnam Cross of Gallantry with Palm. He currently is employed as a computer specialist at the University of Memphis and lives in Memphis, Tennessee, with his wife Sue who graciously allows him to devote many hours to his writing hobby. He was recently commissioned as a Major in the Tennessee State Guard. Tommy has written three non-military books, *"A Million Tomorrows – Memories of the Class of '64," "While Our Hearts Were Young,"* and *"Goodbye to Bob."* He has also edited and published three volumes of memoirs for family and friends. Before contributing and editing this book, he was a writer and editor of three other books in the *"We Were Crewdogs"* series and is a member of The American Author's Association and The Military Writers Society of America.

Rich Vande Vorde

Major, USAF (Ret.) was commissioned a 2Lt in 1971 through the AFROTC Program at University of Minnesota at St. Paul, Minnesota, where he received his MS in engineering. After graduation from Undergraduate Navigator Training and Navigator Bombardier Training (NBT) at Mather AFB, CA, he was assigned to B-52F Combat Crew Training at Castle AFB (for Q-38 training). First assignment was McCoy AFB, Orlando, FL, which promptly closed and then reassigned to B-52Ds at Carswell AFB, TX (for Q-48 training). He flew as a B-52D Navigator and then upgraded to Radar Navigator at Carswell until his 1978 assignment to Mather AFB, CA as an NBT instructor. In 1982 he was assigned again to B-52Gs at Anderson AFB, Guam, where he served on a crew, then as a Wing scheduler. In 1984 he was assigned to K.I. Sawyer AFB, MI where he flew B-52Hs and was a scheduler until 1987, when he was assigned to McConnell AFB, KS as a scheduler for the newly arrived B-1B bombers. He retired four years later in 1991. He and his wife, Cheri, are enjoying retired life now in Paradise, TX, as